INTRODUCTION
TO
INSECT
BEHAVIOR

INTRODUCTION TO INSECT BEHAVIOR

Michael D. Atkins
San Diego State University

Macmillan Publishing Co., Inc.
NEW YORK
Collier Macmillan Publishers
LONDON

Macmillan Publishing Co., Inc.
866 Third Avenue, New York, New York 10022
Collier Macmillan Canada, Ltd.

Library of Congress Cataloging in Publication Data

Atkins, Michael D
  Introduction to insect behavior.

  Bibliography: p.
  Includes index.
  1. Insects—Behavior. I. Title.
QL496.A89   595.7'05   79–15465
ISBN 0–02–304510–8

Printing: 1 2 3 4 5 6 7 8   Year: 0 1 2 3 4 5 6

Preface

Many insects are exhibitionists. They move about and engage in their daily tasks of acquiring food, courting, and reproduction undaunted by the watchful eyes of an observer. Over the span of human life on this planet people have almost certainly been fascinated and entertained by the activities of insects—what child has never stopped to admire a passing butterfly or sat and watched the comings and goings of a column of foraging ants? From such basic observations we have gained considerable insight into the biology of insects, but only recently have we begun to integrate our knowledge of genetics, physiology, and ecology into a more complete view of behavior.

The observation of insect behavior is something every beginning biology student can enjoy, yet even introductory entomology texts devote little space to this intellectually stimulating and practically significant subject. Even more amazing to me is that no modern text has appeared that is devoted entirely to the introduction of all the principal elements of insect behavior. I hope this book will fill that void.

I wrote this text with the purpose of having it serve as the basis for courses in insect behavior and as a supplemental reader for related courses. I have assumed that the students who use it will have been introduced to the insects and have at least some basic knowledge of genetics and physiology. I have presented what I view as being fundamental information about sensory reception, biological rhythms, and basic patterns of behavior before presenting a more detailed view of each of the main biological functions served by behavior and some special topics. Since behavioral processes are so highly integrated, subdividing them into a number of components as I have done may seem artificial. However, this was done to facilitate the development of courses organized in a variety of ways.

This small book is not intended to be a complete coverage of insect behavior—there are much larger books dealing with some of the topics covered here in a single chapter. My aim was to present breadth rather than depth and to select examples that illustrate a point and stimulate a desire to probe a little deeper. The studies cited were a matter of personal preference and in no way slights those pieces of work not mentioned. Furthermore, I accept full responsibility for any error or misinterpretation of fact.

I am indebted to many people who provided guidance, encouragement, and constructive criticism during the preparation of this book. I am especially indebted to Woody Chapman, Biology Editor of the Macmillan Publishing Company, Inc., for recognizing the potential value of this project; and to Fred Delcomyn, of the University of Illinois, and George Evans, of the University of Alberta, Canada, for their helpful comments on the entire manuscript. I also thank my colleagues Theodore Cohn, Edward Huffman, Neil Krekorian, and William Wellington for their ideas, comments, and help with specific chapters.

I thank the following individuals who allowed me to use original material or illustrations in this book: David Bentley, Lincoln Brower, Theodore Cohn, Nicholas Davies, A. F. Dixon, Thomas Eisner, Richard Elzinga, William Evans, Kenneth Hagen, Brian Hocking, D. L. Johnson, Gary Pitman, Carl Rettenmeyer, William Romoser, Herman Spieth, William Stephen, William Wellington, and Edward Wilson. I also thank the following organizations for allowing me to use illustrations from their publications: the American Museum of Natural History; Animal Behavior Annual Reviews, Inc.; the British Ministry of Agriculture, Fisheries, and Food; Cornell University Press; Edward Arnold & Co., London; the Entomological Society of Canada; Harvard University Press; McGraw-Hill Book Company; Melbourne University Press, Melbourne, Australia; the National Research Council of Canada; *Scientific American;* the United States Department of Agriculture; and the University of California Cooperative Extension.

M. D. A.

Contents

INTRODUCTION
TO
INSECT
BEHAVIOR

Introduction

Contemporary animal behavior is a multidisciplinary approach to the search for answers to a variety of fundamental questions about how and why animals do the things they do. The multidisciplinary nature of the subject is understandable in view of the fact that it is often difficult to separate the basic functioning of anatomical components from behavior or the physiology of sensory perception from the basic responses that underlie complex behavioral patterns. Some scientists are interested mainly in the physiological aspects of how stimuli are detected and how responses are initiated. Others are more interested in the patterns of responses, and still others are concerned with the consequences of behavior. Superimposed over this range of basic interests are the conceptual differences that for a long time separated neurophysiologists, ethologists, and animal psychologists.

Ethology has been defined in a variety of ways, none of which is completely satisfactory. However, it is generally accepted that the emphasis of ethology has been the objective study of the biological significance of behavior in the context of nature. In the classical sense, as exemplified by the work of European investigators such as Lorenz, Tinbergen, and von Frisch, ethology was concerned with an evolutionary approach to the study of instinct. On the other hand, classical animal psychology was concerned primarily with the evolution of higher mental processes such as learning and problem solving. Animal psychology tended to be more analytical than ethology and concentrated on experiments conducted in a laboratory setting, more often than not with white rats. Ethology tended to be more descriptive, comparative, and field oriented. These differences are not as apparent today as they were during the 1950s, but differences in approach and emphasis still occur depending on the investigator's underlying discipline. Entomologists, as we will see, deal with a group of animals that behave mainly instinctively, so insect behaviorists have been aligned more with ethology than animal psychology. However, new tools from neurophysiology, genetics, biochemistry, electronics, and statistics have led to a greater unity of purpose and synthesis of thinking among behaviorists studying divergent groups of animals.

In spite of the complexity of behavior, many biologists view it rather simplistically and do not demonstrate much interest in how or why particular patterns of behavior evolved. In entomology, for example, most textbooks devote little or no space to behavior. Volumes have been written about the structural and physiological breakthroughs that have occurred throughout over 300 million years of insect evolution and how these breakthroughs have led to the diversity and success of these fascinating animals. Much less of a general nature has been written about the behavioral adaptations that had to evolve along with the structural changes.

Students of biology often lose sight of the fact that the responses that form the basis of behavior are inherited and, therefore, subject to the same evolutionary processes as inherited structural characteristics; responses that are beneficial or neutral survive, those that are inappropriate are selected against. Clearly, structure and behavior are largely inseparable in the evolutionary process. If a behavioral response already existed that could enhance the usefulness of a new structural modification, the new structure would be more likely to be retained. Conversely, a new structure with potentially significant uses would not persist in the absence of appropriate behavior. Understanding the fundamentals of insect behavior is, therefore, just as vital to the understanding of the success of the insects as is the understanding of their structure.

To the average observer the behavior of insects seems rather stereotyped in that most individuals of the same species tend to do the same things under similar circumstances. For example, if one moves a hand toward a settled housefly, we can predict with reasonable certainty that the fly will take flight. Likewise, if a stone in the garden is raised, we expect the insects beneath it to run quickly toward and under another stone nearby. Sometimes these stereotyped behaviors do not seem to make much sense, as when night-flying moths spend an entire evening fluttering around a porch light. But in each of these and many other examples of insect behavior, we are observing specific genetically programmed responses to stimuli.

Under normal circumstances these basic inherited components of behavior serve a vital purpose and consequently persist as a characteristic of each species. We call these preprogrammed patterns of response **innate behavior** or **instinct**. Innate behavior can be distinguished from learned behavior on the basis that it can be performed with no prior experience. A newly emerged individual can thus respond in a seemingly appropriate manner to both favorable and harmful situations never before encountered. Obviously, the advantage of such a capability to the survival of an organism is immeasurable. However, there are drawbacks, at least for the individual, to behavior that is entirely preprogrammed, since there is no opportunity to develop beneficial alternative strategies in recurring situations. Learning, on the other hand, permits the development of behavioral alternatives that in some situations are more beneficial than the basic innate response.

The extent to which the behavior of a species is dominated by either innate or learned behavior depends on its capacity to learn and, perhaps, its opportunity to learn. The capacity to learn is determined largely by the complexity of the organism's central nervous system, particularly the number of nerve cells in the brain. The major factor governing the opportunity to learn is the time available to experience events to which alternate responses are possible. Consequently, animals like insects, which have a relatively low number (about 100,000) of nerve cells in their brains and generally have quite short lifespans, must depend almost entirely on innate behavior.

Since insects have relatively few nerve cells, most of them must be devoted to a rather fixed set of responses that lead to the accomplishment of those activities that will assure survival and procreation under normal circumstances. These innate responses may lead to the early death of numerous individuals; yet they seem to assure the survival of the population or species. The speed with which insects must respond to many kinds of stimuli led Wigglesworth (1968) to conclude that in the course of their evolution they have sacrificed the refinement of their perception and ability to learn in favor of the ability to react instantaneously to common stimuli.

In addition to not having the longevity conducive to learning, the usefulness of learned behavior among insects is reduced by the complex life cycles that many of them display. The stimuli that are important at different times in the course of development may be quite different, and the various developmental stages often must respond differently to the same stimulus if they are to survive. This is exemplified well by those species that display a high level of structural and ecological divergence between the adult and immature stages. Clearly, the responses

needed for survival may be very different, and the behavior learned
in one stage would be of little value to a subsequent stage that must
live under completely different circumstances. Experiments show,
however, that some insects can and do learn, but when viewed as a
whole we find that learned behavior is not dominant in the insect world.

Humans, on the other hand, with brains made up of billions of
nerve cells, and having comparatively long lives, display mainly learned
behavior. In fact, crying, smiling, and the suckling response of infants
are about the only innate behaviors that can be identified in humans.
Because we have a rather remarkable capacity to learn and solve problems
through experience and practice, many of the statements we make about
the behavior of other organisms reflect the purpose or emotions we
associate with our own actions. These tendencies to attach a purpose
(**teleology**) or human feeling (**anthropomorphism**) to behavior are
tempting pitfalls, especially in the interpretation of insect activities,
because many, particularly social species, display patterns of behavior
that parallel our own. Furthermore, these kinds of interpretation
can obscure the real selective pressures and benefits that led to the
evolution of the innate behavior patterns displayed by many animals.

From the scientists' viewpoint insects are ideal subjects for basic
behavioral studies. As pointed out by Eisner and Wilson (1977), many
of the sensory cells of insects arise directly from identifiable receptor
sites built into the cuticle; these receptors can be manipulated experimentally from the outside with considerable ease. It is possible,
therefore, to break behavior down into basic response components and
determine how tightly or uniformly programmed the behavior is under
different circumstances. The fact that learned behavior does not introduce
an unknown variable makes it possible to interpret more accurately the
adaptive significance of each response.

Beyond the basic acquisition of knowledge of the way organisms
have adapted through evolution, there is a pragmatic reason for studying
insect behavior. The successful management of both beneficial and
harmful species depends on a thorough understanding of all aspects
of their biology. The list of relevant questions that can be asked seems
almost without end, but the answers we obtain could substantially alter
a management practice. For example, we might be able to improve
the efficacy of pollinators by preconditioning them to the fragrance
of the crop we want pollinated. We may be able to explain the success
or failure of applied biological control programs on the basis of the
presence or absence of key stimuli in the host's environment. We may
be able to avoid crop damage caused by an influx of pests through an
understanding of their migratory behavior. We may be able to suppress
a pest population by disrupting a pattern of communication important
to its reproduction.

The chapters that follow will hopefully reveal some of the rather remarkable feats that insects perform instinctively and how understanding their behavior contributes, on one hand, to our overall understanding of their evolution and, on the other, to our ability to make adjustments in our relationship with them. Most insect activities are linked inseparably to some form of stimulus—often several stimuli acting simultaneously. For example, the initiation of walking or flight may be triggered by an interplay of heat and light, and an increase in the intensity of the same stimuli may, in combination with a chemical stimulus, bring about the initiation of feeding. The analysis of these relatively simple components of behavior and how they are initiated requires a careful study of the physiology of the sensory receptors and the central nervous system—subjects that are beyond the scope of this book. Consequently, I will not consider the fundamentals of neural physiology, nor will I discuss how insects walk, fly, ingest their food, or lay their eggs. Instead, I will examine such topics as why they travel and where they travel to, how they select their food, and how they find and identify their mates or oviposition sites.

The first three chapters treat some foundation material including the nature of insect sensory receptors and the stimuli they detect, basic responses and patterns of behavior, and behavioral rhythms. The following seven chapters are devoted to what I consider to be the major biological functions of insect behavior. The final four chapters synthesize the foregoing information through a consideration of eusocial behavior, behavioral ecology, the evolution of behavior, the role of behavior in speciation, and the practical application of behavioral knowledge.

PART I

SOME FUNDAMENTALS

Before one can attempt to understand the functional significance of behavior in terms of the biological functions it serves, one must have a basic understanding of the factors that place constraints upon behavior. These include the organism's capacity to receive information from its internal and external environment, the basic kinds of responses that various stimuli induce, the organism's ability to modify inherited responses as a result of experience, and the sometimes overriding role of patterns of periodicity. These factors form the basis for the three chapters that follow.

Chapter 1 reviews the nature of the cuticular structures that comprise the sensory receptors of insects and describes in a general way the kind of information that can be monitored by them. Chapter 2 reviews the basic response patterns common to all animals and attempts to clarify the sometimes gray area between inherited behavior and learned behavior. Chapter 3 briefly reviews behavioral periodicity and what we understand about the clock mechanisms that tend to govern the behavior of many animals.

CHAPTER 1

Sensory reception

All animals have the ability to obtain information about their environment and respond to that information in an organized manner. Obtaining
the information, **sensory reception**, is the function of a variety of sense
organs or **sensilla**, which incorporate one or more sensory neurones that
transmit the information to the central nervous system for processing.
Any active response to a stimulus results from the activation of a muscle
by one or more nerve impulses from the central nervous system. Since
behavior begins with the reception of one or more signals from the environment, a review of the structure and function of the various sensilla
of insects would seem to be an appropriate starting point for an examination of their behavior.

The sensilla of insects are mostly of cuticular origin arising from
special cells in the epidermis. In a sense they are the windows in the
suit of armor that protects insects from the outside world. Every sense
organ consists of one or more sensory neurons. Most are bipolar neurons;
these have a **dendrite** that runs to a cuticular structure, which protects
the nerve ending and may itself be acted upon in some way by an environmental stimulus, plus an **axon** that transmits the resulting action potential
to the central nervous system. In many sensory structures the dendrite
or a group of dendrites is enclosed in a sheath of cuticular material called
a **scolopale**. Depending on the function of the receptor organ, the dendrite
may end in a **scolopale cap** or the distal end of the scolopale may be open
leaving the dendrite exposed (see Figures 1–1 and 1–5).

Each type of sensillum has evolved effectively to detect a specific kind
of energy or stimulus. One group, called **mechanoreceptors**, receives
mechanical energy generated by physical forces such as touch, pressure,
or vibrations through the substrate or the surrounding air or water (this
would include sound reception). Another group detects the chemical
energy of various molecules (**chemoreceptors**), and a third group detects
electromagnetic energy in the form of heat (**thermoreceptors**), light
(**photoreceptors**), and perhaps electromagnetic fields. Regardless of the
form of energy that excites a sensory neuron, it is transduced into the
electrical energy of the nerve impulse transmitted to the central nervous

system. The nature of each stimulus is decoded by the central nervous system on the basis of the type and location of the sensory receptor that initiated the signal and where in the nervous system the information is processed. The regions of the front portion of the brain known as the mushroom bodies or **corpora pedunculata** are believed to be the sites where signals arriving simultaneously from a number of different receptor areas are integrated into complex behavior patterns. These centers of the brain are also thought to be involved in learning and are largest among the social Hymenoptera.

MECHANORECEPTION

Sensory structures capable of detecting mechanical forces occur all over the inside and outside surfaces of the insect integument. The most common and obvious of these structures are hairlike cuticular outgrowths called **setae** or **trichoid sensilla**. These sensory organs consist of a projection of cuticle articulated with the body in a membranous socket. The seta itself is produced by a specialized epidermal cell called a **trichogen**; its socket is produced by an adjacent cell called a **tormagen**.

Trichoid sensilla are involved in the reception of a variety of stimuli and vary in design accordingly. Those concerned with mechanoreception usually have a single sensory neuron. The dendrite of these sensilla commonly terminates in the scolopale cap, which is inserted into the cuticle of one edge of the base of the seta (Figure 1–1). Any movement of the hair due to a mechanical force produces a distortion of the basal membrane and the base of the seta, which is detected by the sensory dendrite. Some trichoid sensilla excite the associated neuron when deflected in any direction, but others only excite the dendrite when bent in some particular direction.

Most trichoid sensilla are excited only when being bent or straightened, but some are excited all of the time the seta is out of its normal position. These two functional types of hairs along with other forms of mechanoreceptors that function in a similar way are called **phasic** and **tonic** receptors, respectively. Dethier (1968) applied the terms **velocity sensitive** and **pressure sensitive** on the basis that the former respond rapidly to forces, like velocity, that change the degree to which the hairs are deflected, whereas the latter respond to the static state of deflection by sending a steady flow of impulses to the central nervous system.

Trichoid sensilla are located on almost all parts of the body, but they are not distributed evenly. Frequently, they occur in dense patches called hair beds and are more numerous on those parts of the body such as the mouthparts, antennae, and legs that come in contact with each other or other surfaces. Some phasic sensilla can be extremely sensitive and detect

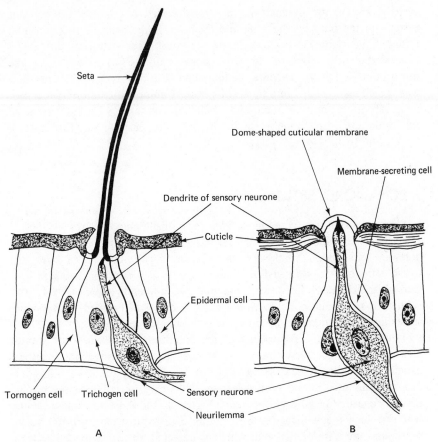

Seta

Dome-shaped cuticular membrane

Membrane-secreting cell

Dendrite of sensory neurone

Cuticle

Epidermal cell

Tormogen cell   Trichogen cell   Sensory neurone

Neurilemma

A   B

Figure 1–1. Structural details of two common mechanoreceptors of insects. (A) Hair or trichoid sensillum. (B) Dome sensillum. (Redrawn from R. E. Snodgrass, *Principles of Insect Morphology*, McGraw-Hill Book Company, New York, 1935.)

even the slightest of air currents produced by the movement of a nearby object. Such sensitivity can aid immeasurably in the initiation of escape reponses. A number of insects have tactile hairs on the front of the head that detect the flow of air during flight and may provide a means of determining flight speed; some insects cease flying when these hairs are not adequately stimulated.

Patches of tonic trichoid sensilla located between the joints of the legs and other adjacent body parts provide an insect with information about its position in its environment and the position of different parts of the body relative to one another; this is referred to as **proprioreception**. Ants have sensory setae between the head and thorax (Figure 1–2) that detect movements of the head (Markl, 1962). In a similar way hair beds

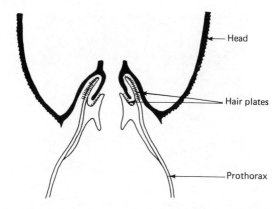

Figure 1–2. Transverse section through the head and prothorax of the ant *Formica polyctena*, showing the position of the sensory hair beds that detect the movement of the head. (Redrawn from H. Markl, *Z. Vergl. Physiol.*, **44:** 475–569, 1962.)

on the anterior portion of the prothorax inform the mantis about the position of its head as it turns to follow a potential prey. The hair beds are stimulated by increased pressure on the side of the body to which the head has turned, and the nervous impulses transmitted allow the central nervous system to coordinate the direction of the grasping extension of the grasping (raptorial) forelegs.

Hair beds also serve as a means of orienting to gravity. Patches of trichoid sensilla between the head and thorax, and between the thorax and abdomen, serve this purpose in the honeybee *Apis mellifera*. The perception of gravitational pull is of particular importance in this species because gravity is used as the reference for communicating the position of the sun when the bee dance is performed on a vertical surface. Since the center of gravity of the head is ventral to its point of articulation with the thorax, when the bee moves its body upward the pull of gravity deflects the head downward, thereby increasing the pressure on the ventral cervical hair beds (Lindauer, 1961).

An additional contribution to proprioception is made by **campaniform sensilla**, which consist of thin cuticular domes, each with a single dendrite, usually enclosed in a scolopale, inserted into it. These structures occur in groups concentrated at the leg joints and at the bases of the wings, or the halteres of flies. They respond to pressure and elastic movements of the cuticle and are concerned primarily with movements of the legs and wings.

Another important group of mechanoreceptors are known as **chordotonal organs**. These structures consist of one or more bipolar neurons, the dendrites of which are ensheathed in a scolopale and attached to the cuticle at one or both ends; the neurons lie on the inner surface of the

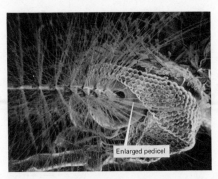

Enlarged pedicel

Figure 1–3. Scanning electron micrograph of an antenna of a male mosquito showing the long hairs arising from the flagellum and the enlarged pedicel that houses Johnston's organ. (Photograph by M. D. Atkins.)

cuticle and lack outwardly visible structures. Chordotonal organs are much less abundant than trichoid sensilla and may have rather specific patterns of distribution in different orders of insects (Debaisieux, 1938). Chordotonal organs are sensitive to vibrations of the substratum and to vibrations through the surrounding medium, including airborne sounds.

Johnston's organ is a complex chordotonal organ that lies in the second segment of the antenna of the adult form and some immature forms of all true insects. It is most highly developed among the males of mosquitoes and midges, and their pedicels are enlarged accordingly (Figure 1–3). Johnston's organ of mosquitoes consists of two rings of scolopidia that connect the inner surface of the pedicel and a cuticular extension of a basal plate that arises from the lower segment of the antennal flagellum. There are three additional single scolopidia, which extend from the first antennal segment to the base of the flagellum (Figure 1–4). In mosquitoes and a variety of other insects, Johnston's organ serves for sound perception. The densely plumose antennae of the male mosquito vibrates when impinged upon by sound waves of a frequency close to that of the wing beat frequency of the female.

In the fly, *Calliphora,* Johnston's organ serves as a flight speed indicator by detecting the degree to which the third antennal segment rotates relative to the pedicel as a result of air pushing on the arista. Although the sensilla of the organ are mainly phasic, they are stimulated by vibrations of the antennae caused by the moving air (Chapman, 1969). In whirligig beetles, Johnston's organ perceives antennal displacement caused by ripples traveling across the surface of the water instead of sound. By these means, the beetles can determine the location of other individuals and, by detecting "echoes" of their own ripples, can avoid collisions with large objects in the water.

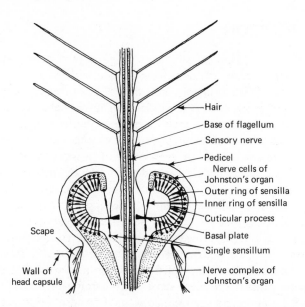

Figure 1–4. Diagrammatic representation of a transverse cross section through the basal portion of a mosquito antenna showing the structure of Johnston's organ. (Adapted from R. F. Chapman, *The Insects: Structure and Function,* American Elsevier Publishing Co., Inc., New York, 1969.)

Johnston's organ may also play a role in orientation, since it should detect any deflection of the antennal flagellum due to gravity. In the back swimmer *Notonecta,* an air bubble that extends between the head and the antennae when the insect is in its normal swimming position deflects the antennae away from the head. If the insect is turned over, Johnston's organ responds to the change in position of the antennae.

Members of the orders Orthoptera, Hemiptera, and Lepidoptera have special chordotonal hearing organs called **tympanal organs**. These organs, found on the legs, thorax, or abdomen, depending on the family, consist of a membrane or tympanum stretched across an air space or tympanic cavity so that it is free to vibrate at the frequency of the sound waves that impinge upon it. From 2 to 1500 chordotonal sensilla are associated with the tympanic membrane and are excited by the sound that makes it vibrate. How the vibratory stimulation of the sensory cells is transduced is not known, but they are capable of detecting sounds over a wide range of frequencies, depending on the species. The principal function of tympanic organs is to detect potential predators or the stridulatory sounds produced by members of their own kind.

A more detailed account of mechanoreception was presented by McIver (1975). The physiology of mechanoreception is covered extensively by Schwartzkopff (1973).

CHEMORECEPTION

The detection of chemical stimuli is of great importance to insects in that it provides a means of identifying a wide variety of substances both harmful and beneficial. Probably, no other aspect of insect sensory reception has been of more interest to students of insect behavior. The chemosensory processes exercise a major degree of control over reproductive behavior, insect-host relationships, habitat location and selection, plus the integration of many aspects of presocial and eusocial behavior. Chemoreception is reviewed in considerable detail by Hodgson (1973).

Chemoreceptors are of considerable variety and are widespread, although concentrated on the antennae, mouthparts, and tarsi. The one feature they share is that the sensory dendrites are exposed through small openings in the cuticle. However, the chemical concentration detected varies according to the number of sensory neurons that a single receptor contains. Some entomologists feel that insects have what might be called a **common** or **general chemical sense**; this permits them not to respond to substances they encounter frequently unless those substances occur in potentially dangerous concentrations, in which case an avoidance reaction may be beneficial. Although no specific receptors have been identified that provide the general chemical sense, such receptors are believed to be of a type only responsive to high concentrations by involving few sensory neurons with limited exposure (Figure 1–5A).

In addition to their general chemical sense, insects are often said to have a sense of smell (**olfaction**) and a sense of taste (**gustation**). However, this is a dichotomy that is not justified on the basis of how the perception is accomplished, as all chemicals are perceived in basically the same manner. The chemical molecules must come into direct contact with the sensory dendrites of the receptor organ in order that the "potential energy existing in the mutual attraction and repulsion of the particles making up the atoms" can be detected (Dethier, 1963). Although statements made by behaviorists imply that insects can detect chemicals both on contact and at great distances, they are really referring to the ability to locate the source of the chemicals. Since molecules must impinge upon a receptor neuron before they can be sensed, insects can only locate distant sources if they can follow a gradient of the chemical to its origin. Generally, olfaction refers to the perception of chemicals in a vapor state, whereas gustation is the perception of solids or liquids.

The fact that the chemical molecules must deliver their energy directly to the sensory neurons requires a diffrent form of receptor organ than those used to detect mechanical stimuli; for mechanical stimuli the sensory neurons are attached to a structure that responds to vibrations or changes in pressure. Chemoreceptors are generally characterized by having very fine nerve endings exposed to the environment through minute openings

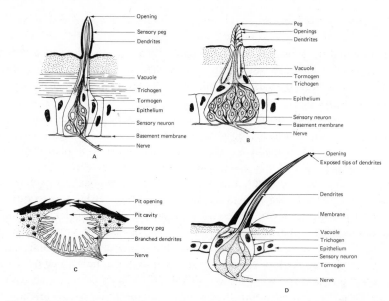

Figure 1–5. Diagrams of several different kinds of insect chemoreceptors. (A) A thick-walled peg of the type involved in general chemical reception. (B) A thin-walled peg of the type involved in olfaction. (C) A typical olfactory pit containing a number of sensory pegs. (D) A modified hair sensillum of the type involved in gustation. (After E. H. Slifer, J. J. Prestage, and H. W. Beams, *J. Morph.*, **105**: 145–191, 1959.)

in the cuticle. Those receptors that detect chemicals in a gaseous state in relatively low concentrations normally have a large number of sensory neurons (Figure 1–5B and C). Each neuron responds either to a range of compounds or to some special compound of importance. Those that detect substances in a high concentration, in solution or a liquid state, have a smaller number of sensory neurons (Figure 1–5D). Although the mechanism of detection is the same in each case, the olfactory receptors enable the insect to detect chemicals in the air at some distance from their source, whereas the gustatory receptors are better designed for the identification of chemicals at their source.

Olfactory receptors are most commonly found on the antennae but are also found in large numbers on the mouth parts, particularly the palps, and the ovipositor. Gustatory receptors are extremely abundant on the mouth parts, as one might expect, but, more surprisingly, they occur in large numbers on the tarsi of some insects. Although the response to chemical stimulation is now measured by a sophisticated apparatus that detects changes in the electrical potential of nerves carrying the impulses from groups of chemoreceptors, it is quite easy to demonstrate the presence and function of these organs without such equipment. For example, insects often display an orientation to certain food or to mates that can be observed in the laboratory. However, if their antennae are removed or coated with an impermeable substance, they no longer display the predicted orientation. The presence of taste receptors on the tarsi of a honeybee can be demonstrated by bringing the tarsi into contact with a sugar solution and noting that the bee will immediately extend its proboscis in search of the food.

For a response to a chemical to occur, a number of molecules must impinge upon the dendrites of receptors for that class of compound. In cases where contact sensilla are exposed to concentrations of a chemical, the substance must produce a depolarization of the dendrite membrane to initiate an impulse; however, the nature of the mechanism by which this is brought about is not known (Chapman, 1969). In the detection of chemicals in a gaseous state, the concentration that will trigger a response is determined to some extent by the number of sensilla present, by the number of sensory neurons in each sensillum, and the degree of branching of the dendrites. For those chemicals that must be detected in low concentrations and for which a low threshold for a response is necessary, the number of exposed dendrite endings is highest. In many insects the surface area that supports chemoreceptors is expanded. This is particularly obvious from the elaborate configurations of insect antennae. For example, the terminal club of a scarab beetle antenna consists of a number of receptor-bearing lamellae that are spread apart when the beetle is searching for food or a mate.

HYGRORECEPTION

The perception of moisture in the air, **hygroreception**, is not very well understood for insects, even though there is ample behavioral evidence to support it. The detection of water is a form of chemoreception, in the sense that water is a chemical compound, but, at least in some cases, it may be perceived more like a mechanical stimulus. Many animal hairs are known to straighten and curl in relation to the wetness of the environment, which is why horsehair is used as the sensitive element in hygrographs. If insect hairs respond to moisture in a similar manner, the distribution of tension across the membrane of the hair's socket would change and thereby stimulate the associated sensory neuron. The relative humidity also affects the drying power of the air and, consequently, the amount of evaporational cooling that may be detected by thermoreceptors. In the human louse, *Pediculus humanus,* which is particularly sensitive to changes in the moisture content of the air, the antennae bear tufts of hair that absorb water relative to the amount present in the adjacent atmosphere. Recent studies suggest that antennal receptors that specifically detect the moisture content of the air may occur widely in the Insecta (Loftus, 1976; Harbach and Larson, 1977; Yokohari, 1978).

THERMORECEPTION

Because insects are cold-blooded animals, almost all of their activities and functions are regulated or affected by the temperature of their environment. It is not surprising, then, that insects are often able to detect rather small changes in temperature and to display a variety of heat responses, including a tendency to aggregate in a particular zone of a temperature gradient. In addition to displaying what we might call a general temperature sense, some insects use heat detection for specific purposes, such as the location of hosts. Many parasites of warm-blooded animals apparently orient to their hosts by following the temperature gradient surrounding the host's body. For example, some biting flies locate their hosts by responding to the temperature gradient and/or convection currents generated by their hosts. The bedbug, *Cimex* (Hemiptera), displays a strong general temperature sense and will leave a feverish person in favor of a bedfellow with a normal body temperature.

Despite all of the observed evidence for heat sensitivity among insects, we have little well-substantiated information about the nature of heat receptors. In the blood-sucking bug, *Rhodnius,* groups of thick-walled setae on the antennae are presumed to be heat sensilla. One of the most exciting recent additions to our understanding of sensory reception by insects (Evans, 1975) involves the discovery of special infrared radiation detectors on the underside of the mesothorax of the buprestid beetle, *Melanophila* (Figure 1–6). These beetles oviposit in the still hot bark of trees scorched by forest fires. Their infrared detection devices enable them to scan several miles of countryside while flying in search of trees recently damaged by fire.

PHOTORECEPTION

The detection of that part of the electromagnetic spectrum referred to as light is called **photoreception**. We tend to think of light reception as vision because the stimulation of our light-sensitive organs usually results in the formation of an image. However, many animals are unable to perceive images and respond only to the presence or absence of a light stimulus. The insects as a group are able to detect light versus darkness, as well as photoperiod, light intensity, the plane of polarization, movement, form, patterns, and some colors. To which of these light-related forms

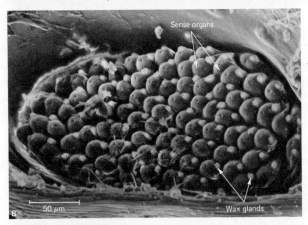

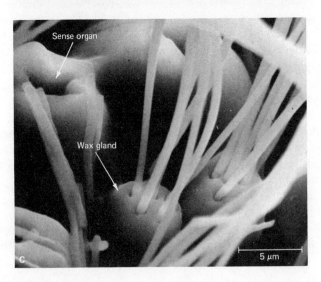

of environmental information an individual insect can respond depends upon the type of receptors it has.

Some insects do not have identifiable light receptors but display a **dermal photo response** to light striking or penetrating the cuticle. The maggots of some higher Diptera have groups of photosensitive cells on the anterior of their bodies that cause a simple orientation and movement away from the source of stimulation. However, most insects have well-developed photoreceptors in the form of simple eyes and compound eyes.

Most adult insects have a pair of compound eyes, one eye on each side of the head, that tend to bulge outward and provide a wide field of vision. However, the organs have been secondarily lost in the adults of some parasitic, sedentary, and cave-dwelling forms for which normal vision is no longer important. Compound eyes are also a characteristic of the immature stages of most species that have incomplete metamorphosis but are never present among the larvae of species with complete metamorphosis.

Each compound eye consists of a group of structural units called **ommatidia**. The number of ommatidia that make up the compound eye varies greatly from species to species and is believed to be correlated directly with visual acuity. The compound eyes of the thysanuran *Lepisma,* which is wingless, consist of only about a dozen ommatidia more or less separated from one another, whereas the large compound eyes of active fliers like the dragonfly have up to 10,000 ommatidia, which are packed tightly together, giving the surface of the eye a honeycomblike appearance. In many insects, specific areas of the eye, specialized to provide greater visual acuity, have more densely packed ommatidia than other areas.

Each ommatidium (Figure 1–7) consists of an outer cuticular lens or **cornea**, which is transparent and more or less biconvex. The lenses combine to form the surface of the compound eye, each one making up an area known as a **facet**. Immediately beneath the cornea are four cells, which together form the **crystalline cone**, surrounded by a group of corneagen or **primary pigment cells**. The light-gathering components just described overlie the receptor apparatus. The receptor portion of the ommatidium consists of 6 to 12 sensory **retinular cells** (Horridge and Giddings, 1971), each of which is connected to the optic nerve via its axon. The retinular cells produce a central rodlike light-sensitive unit called a **rhabdom**. The rhabdom contains visual pigments, **rhodopsins**, at least one of which has been identified (Paulsen and Schwemer, 1972). The response of the

---

Figure 1–6. Scanning electron micrographs of an infrared sensory pit of the buprestid beetle *Melanophila acuminata*. (A) An area of the pit showing the mat of wax filaments covering the sense organs. (B) The sensory pit with the wax filaments removed. (C) A sense organ and adjacent wax glands extruding wax filaments. (Courtesy of William G. Evans.)

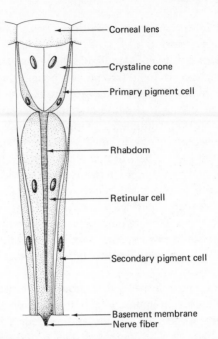

Corneal lens

Crystaline cone

Primary pigment cell

Rhabdom

Retinular cell

Secondary pigment cell

Basement membrane
Nerve fiber

Figure 1–7. Transverse cross section through a single light-sensitive unit called an ommatidium.

visual pigment to increased illumination is a conformational change in its molecular structure that results in an increased depolarization potential and a correspondingly greater output of the sensory neurons to the brain (Goldsmith and Bernard, 1972). The rhabdom and retinular cells of each ommatidium are surrounded by **secondary pigment cells,** which variously restrict the light that can enter through adjacent lenses. Each ommatidium perceives that part of the insect's visual field from which light stimulates its rhabdom. The individual images received are then combined by the brain into a **mosaic image,** which often has been likened to our viewing the environment through a bundle of drinking straws. This is not an accurate simile, however, as an insect would perceive overlapping images from adjacent ommatidia.

The area of the environment and the clarity of the image detected by each ommatidium is dependent upon the angle of the incoming light waves, and it is useful to recognize two basic types of compound eyes that are different in this respect. **Photopic** or **apposition eyes** (Figure 1–8A) have rhabdoms that are long, thin, and extend from the basement membrane to the crystalline cone. In many species there is little or no movement of the secondary pigment, and each rhabdom is stimulated only by light entering the lens directly above it. As one might expect, eyes of this type are commonly found among species active in bright light. **Scotopic** or **superposition eyes** (Figure 1–8B and C) have short, thick rhabdoms,

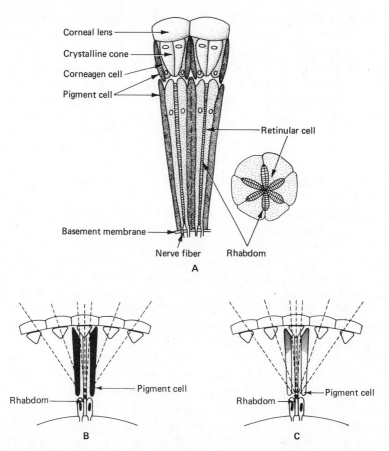

Figure 1–8. Various types of compound eye. (A) An apposition eye.
(B) A light-adapted superposition eye showing restricted light penetration
due to distribution of the pigment in the secondary pigment cells. (C) A
dark-adapted superposition eye showing outward migration of pigment in
the secondary pigment cells allowing access to light from adjacent lenses.
(Courtesy of William S. Romoser.)

separated from the cone by some distance, and there is considerable
longitudinal movement of the secondary pigment that isolates the ommatidia to varying degrees in relation to the light intensity. At higher intensities, the pigment is distributed so as to isolate the ommatidia (Figure
1–8B), in which case the eye functions as an apposition eye and is said to
be light adapted. At low light intensities, the pigment moves to the outward part of the secondary pigment cells so that some light passes obliquely
from ommatidium to ommatidium without being absorbed (Figure 1–8C),
and the eye is said to be dark adapted; this is of considerable advantage to insects that are active during times when little light is available. Although there
is not complete agreement, some workers have theorized that photopic

eyes or light-adapted scotopic eyes form more highly resolved images. Because more light from a limited portion of the visual field strikes each individual rhabdom, an over-all image composed of fairly well delineated, intense subimages is formed. On the other hand, the dark-adapted scotopic eye gives up some of the power of resolution in order that sufficient light can be received by each rhabdom to induce stimulation; however, there is considerable overlap of the images received by adjacent ommatidia.

The position of the eyes on the head and the curvature of their surfaces provide insects with very extensive visual fields. In some, the visual field extends through 240° horizontally and 360° vertically, and the fields of the two eyes overlap extensively to provide binocular vision above, below, and in front of the head.

Because our eyes are completely different structurally from the eyes of insects, we cannot develop a device that would enable us to see what an insect sees. Consequently, we develop our understanding of the insect's visual perception from an interpretation of behavioral evidence and the results of experiments. Much of the work related to insect vision has been conducted with honeybees because they can be trained to repeat behavior patterns through the offering of a reward in the form of sugar and water. Experiments have shown that bees can be trained to associate food with certain types of pattern that contrast with a uniform background color. Although they seem unable to distinguish between shapes such as triangles, squares, and circles that we differentiate readily, they can distinguish solid forms from figures broken up into contrasting pieces. Studies have also shown that bees respond more readily to broken patterns that produce a high frequency of change in retinal stimulation. Wigglesworth (1965) suggested that the divided nature of flowers and also flowers moving in the wind produce flickering images that the bees associate with the presence of nectar. Hence their ready response to a broken pattern that produces a flicker effect is very useful.

Insects are also capable of perceiving color. The visual pigments of insects absorb different wavelengths of light unequally. Consequently, their light receptors are stimulated by some wavelengths and not others. Experimental evidence reviewed by Goldsmith and Bernard (1972) suggests that insects may have ommatidia with different wavelength sensitivities and/or ommatidia with visual pigments with several maximal wavelength sensitivities. Although there is considerable variation in terms of the wavelengths detected by different insects, as a group they are sensitive to wavelengths from about 240 (ultraviolet) to 650 (yellow-orange) nanometers compared to the human eye, which is sensitive to wavelengths from about 400 (blue-violet) to 800 (red) nanometers. Insects are particularly sensitive to the ultraviolet and blue-green part of the spectrum. Studies on the color perception of honeybees have shown that they respond more strongly to blue and violet than to yellow-green and yellow, and are unable to distinguish red from black because both are perceived as noncolors.

Many insects are able to detect ultraviolet light, and this part of the spectrum plays several important roles in insect behavior, which will be discussed in later chapters. Ultraviolet and near ultraviolet light have been shown to stimulate directional light responses in both diurnal and nocturnal species. The major source of these wavelengths is the open sky, but they also penetrate cloud cover and form a major component of polarized light, thereby enabling insects to determine the position of the sun even when it is obscured from their view.

The detection of movement is also of considerable importance to insects, and the structure of the compound eyes appears to be better adapted to detecting movement than images. As the eyes are composed of numerous, separate sensory units, movement can be detected as a source of stimulation that travels across the eye and activates a series of ommatidia in sequence. Rapid, repeated movements, however, may not be detected because the sensory units require time to recover between stimuli. The recovery time required between stimuli varies from species to species but appears to be shorter for rapidly flying insects—a rather fascinating adaptation for gaining visual information from rapidly passing terrain. Some insects, such as mosquitoes, will continue to fly on a fixed mount in still air only if a cross-striped pattern is continually passed beneath them from front to rear. If the direction of movement of the pattern is reversed, they will stop beating their wings even if a stream of air is directed backward over their heads. This suggests that the direction of pattern movement seems to provide an indication of flight speed relative to the ground.

Many insects must also be able to judge distance for a variety of reasons, but this is particularly important for insects that capture prey. Distance perception requires simultaneous stimulation of groups of ommatidia located in the same region of the two eyes. This can be demonstrated easily by damaging or masking one eye. However, insect eyes are fixed relative to each other; thus the insect cannot determine distance by converging on a fixed point as is possible in vertebrates. The distance to an object is determined by its position in the visual field of the two eyes and the point of intersection of projected axes of the simultaneously stimulated ommatidia. Maldonado and Barros-Pita (1970) have shown that in the praying mantis the image of a prey must fall into the region of high resolution, called the **fovea**, before the mantis will strike.

Many adult insects have **dorsal ocelli** in addition to compound eyes, whereas the larvae of insects that undergo complete metamorphosis never have compound eyes and usually have **lateral ocelli**, called **stemmata**, instead. The structure of stemmata is quite variable. Some, like those of caterpillars, are rather similar to a single ommatidium, as previously

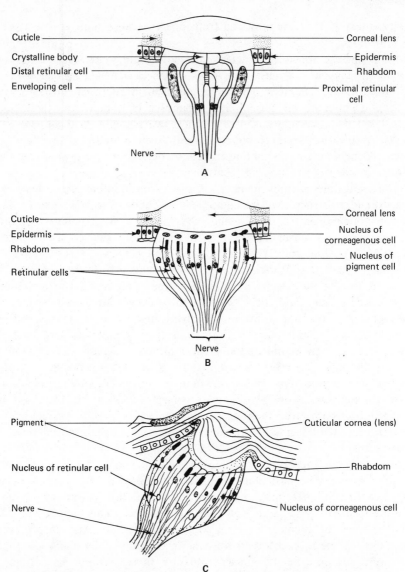

Figure 1–9. Different types of simple eyes. (A) Lateral ocellus of a caterpillar showing single centrally positioned rhabdom. (B) Lateral ocellus of a sawfly showing absence of the crystalline body and numerous rhabdoms. (C) The dorsal ocellus of a hemipteran.

described; each consists of a cornea, crystalline cone, and a group of retinal cells that form a single rhabdom (Figure 1–9A). In other insects, there is a single cornea that overlies a group of retinal cells and several rhabdoms (Figure 1–9B). Regardless of their structure, stemmata function like the ommatidia of compound eyes and are believed to detect images, form, color, and movement. Clearly, a small number of sensory units would not permit the formation of an elaborate mosaic image, but insects that have these light-sensitive organs are believed to be capable of scanning their environment by swinging their heads back and forth.

Dorsal ocelli are similar in structure to the second type of stemmata mentioned previously (Figure 1–9C). Typically, there are three separate ocelli located in a triangular pattern on the anterior part of the vertex of the head; however, there may be two or none. Consequently, dorsal ocelli do not detect form as far as we know. Their main function is to detect changes in light intensity, possibly to provide a general stimulation of the nervous system that makes the insect more responsive to other environmental stimuli (Dethier, 1963). Goodman (1970) provides a detailed review of insect ocelli.

CHAPTER 2

Basic responses and
patterns of behavior

In the previous chapter we examined the important kinds of environmental stimuli that insects are capable of detecting and the anatomy of the principal kinds of structures that mediate their reception. In this chapter we shall review the ways in which insects respond to these stimuli, how sequences or patterns of response can serve vital functions, and how such patterns may even be modified somewhat by experience.

Any basic response that can be performed spontaneously the first time an appropriate situation presents itself is clearly **unlearned** and is referred to as **innate behavior**. Often, a number of inherited responses are combined into complex patterns of behavior that, although having the appearance of a learned reaction or procedure, are not the result of prior experience. On the other hand, a simple response may be modified by experience and thereby qualify as **learned behavior**. Consequently, in the absence of experimental evidence, the distinction between innate and learned behavior can be difficult.

INNATE BEHAVIOR

Innate or interited behavior consists of a wide range of responses; some involve movement and some do not. These responses are most commonly classified as **reflexes, kineses, taxes,** and **transverse orientations.**

REFLEXES

The simplest form of innate behavior is called a reflex. The rapid involuntary withdrawal of your finger from a hot object is a typical reflex reaction. This simple form of response involves a receptor organ complete with at least one sensory neuron as well as a motor neuron connected to an effector organ, in this case a muscle. When dendrites of the sensory neuron are excited, an impulse flows along the nerve to the end of its axon. The impulse is then transmitted across a synapse either to an association neuron and then across another synapse to a motor neuron or directly to a motor neuron; the axon of the motor neuron terminates in a muscle fiber. This simple pathway of stimulus reception and response, called a **reflex arc**, is presented diagrammatically in Figure 2–1. Some physiologists divide reflexes into two functional groups. **Phasic reflexes** are rapid, short-lived responses, involved in the rapid movement of one part of the body or the entire body. For example, many insects respond to a threat by suddenly taking flight, a reflex also known as an **avoidance reaction**. **Tonic reflexes**, on the other hand, are slow, long-lived responses that maintain posture, the position of the body in space, muscle tone, and so forth.

Reflexes vary in complexity and usually occur in sequences or patterns coordinated within the ganglia of the central nervous system. The most apparent reflexes are those involved in the orientation of the body in space or relative to specific sources of stimulation. Jander (1963) defined orientation as "the capacity and activity of controlling location and attitude in space and time with the help of external and internal references, i.e., stimuli." The maintenance of the position of the body is often called **primary orientation**, whereas directional responses to stimuli are called **secondary orientation**. Most studies of individual behavior are concerned with secondary orientation. Such studies reveal a great deal about the stimuli that dominate behavior but may be complicated by the current physiological state of the individual; the same individual may orient toward a stimulus at one time, but subsequently reverse its orientation, or not orient at all.

The classification of the orientation of animals in motion has become complex, and unfortunately the use of an extensive technical terminology varies from one behaviorist to another. Although much new information

Figure 2–1. A simple reflex arc of the type involving an association neuron.

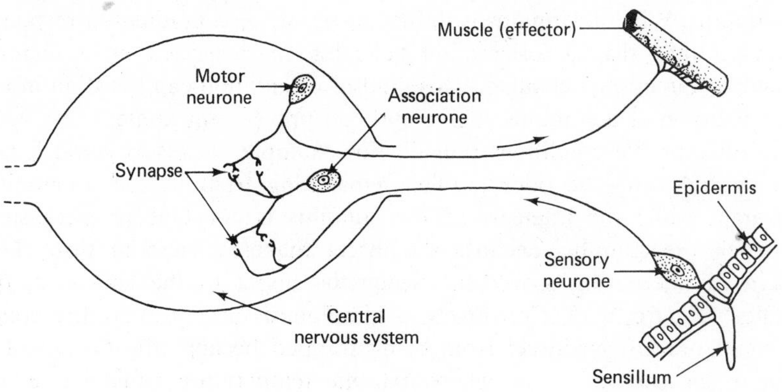

concerning insect orientation has been gained in recent years, the reviews by Fraenkel and Gunn (1961) and Markl and Lindauer (1965) are among the most useful. In addition to primary and secondary orientation, Fraenkel and Gunn (1961) describe three basic kinds of oriented behavior—**kineses, taxes,** and **transverse orientations**. These three mechanisms are exceedingly important in directing the movement of insects from inhospitable to hospitable environments, and to the requisites for life (food, mates, shelter, etc.).

KINESES

Kineses are essentially random or undirected locomotor reactions, the intensity of which varies with the intensity of the initiating stimulus; no particular orientation of the long axis of the body results. The simplest type of kinesis, called an **orthokinesis** (*orthos,* straight; *kine,* movement), is basically a movement response to a stimulus. For example, an insect that is inactive in total darkness may start to stir when exposed to a low level of light intensity. As the light intensity is increased slowly from darkness, an intensity is eventually reached at which the insect becomes active; this intensity is the threshold for the light stimulus for the individual being observed under the prevailing conditions. If the light intensity (the strength of the stimulus) is increased further, the rate of movement will increase as well, until some upper stimulus threshold is reached and activity ceases. In other words, there is a direct or straightforward relationship between the intensity of the stimulus and the intensity of the response, between a lower and an upper threshold.

Whereas an orthokinesis may lead to aggregations of individuals that have simply entered a zone of low-level stimulation and become inactive, aggregations are more likely to result from **klinokineses** (*klinein,* bend; *kine,* movement). Klinokinetic responses are characterized by a change in the frequency of random turns in relation to changes in the strength of the stimulus. When a stimulus is diffuse or occurs as a gradient, a responding insect may display orientation behavior characterized by a number of random directional changes. This kind of orientation can assist an insect in the location of a stimulus source and can prevent entrapment in a zone where unfavorable conditions prevail. For example, an insect could locate food by following the odor gradient emanating from it; the orientation is straight while the intensity of the stimulus is constant or increasing, but when the stimulus declines the insect makes a random turn. This pattern of alternating movements leads the insect to the source of the stimulus (Figure 2–2). Conversely, a klinokinesis in a temperature gradient could prevent an insect from being trapped because the turns would result in its leaving a zone where extreme temperature could cause immobilization.

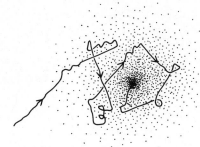

Figure 2–2. The hypothetical track of an insect orienting toward the center of a concentration gradient such as for an odor stimulus. The track remains straight as long as the strength of the stimulus remains constant or increases, but random turns are made whenever the strength of the stimulus declines. (Adapted from V. B. Wigglesworth, *The Life of Insects,* The New Amsterdam Library, Inc., New York, 1964.)

TAXES

Taxes are directional movements toward (positive) or away from (negative) a source of stimulation, with the orientation along a line that runs through the source and the long axis of the animal's body. The sensory system involved must be capable of determining the direction from which the stimulus comes, and the insect must be able to correct angular deviations from the true course. This is achieved by a turning tendency that increases in magnitude directly with increased angular deviation. A taxis, therefore, provides an efficient means of orientation.

Taxes (*taxi,* formal arrangement) in general are classified according to the type of stimulus that initiates them [e.g., phototaxis (light), skototaxis (darkness), geotaxis (gravity), amenotaxis (air current), rheotaxis (water current), thigmotaxis (contact), and so on]. They are also classified according to the manner in which the animal deals with differences in the strengths of stimuli, stimuli from more than one source, or mixtures of stimuli.

Klinotaxes are orientation reactions to or away from a source of stimulation involving regular alternating deviations as a necessary component of the orientation. An insect displaying a klinotactic orientation swings the anterior portion of its body back and forth across the field of stimulation because the receptor(s) are not equally accessible to multi-directional stimulation but move more or less directly to or away from the source of stimulation (Figure 2–3).

Tropotaxes (*tropo,* change) are characterized by a path that is straight and directly toward or away from the source of stimulation by the use of paired receptors, one of which is located on each side of the animal's

Figure 2–3. Orientation of a caterpillar toward a single light source achieved by swinging the head from side to side, thereby balancing the stimulation received by the ocelli on each side of the head.

body. Insects responding in this way to light have been described as moving "as if spitted on a light ray." The sensory receptors must be arranged so they are not usually stimulated equally. In a light response, for example, if one eye is stimulated more than the other, the insect will turn toward the more intensely stimulated eye until a balance between the impulses from the two eyes is established in the central nervous system. A phototropotaxis can be demonstrated experimentally in several ways. If a photopositive beetle is placed on a platform in front of a point source of light, it will soon begin to walk in a straight line toward the light. When the platform is slowly rotated to the right, the beetle will make a continuous compensatory turn to the left (Figure 2–4). If one eye is masked, the beetle will also continue to turn toward the light until the uncovered eye can no longer see it. If the unilaterally blinded beetle is then placed in a white arena lit from above, the turning will continue toward the unmasked eye.

Another characteristic of a tropotactic response is a balanced orienta-

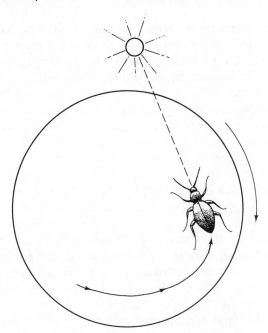

Figure 2–4. The curved track of a photopositive beetle orienting to a fixed single-light source while walking on a slowly rotating platform.

tion between two equal sources of stimulation. If an animal so stimulated goes to one of the sources instead of between them, the response is called a **teleotaxis** (*telo,* end or objective). In a two-light experiment, for example, the insect would clearly orient toward one light at a time although it might switch its orientation to the other light periodically so that it follows a zigzag course. This suggests a central inhibition of response to one of the sources of stimulation. Bees have been shown to be capable of both tropotactic and telotactic orientation in two-light experiments, but there are times when choosing one of two stimuli, rather than going between them, would be clearly beneficial. As Fraenkel and Gunn (1961) suggest, it would serve little purpose for a bee visually orienting to flowers to pass between them.

TRANSVERSE ORIENTATIONS

Transverse orientations result in an alignment of the body at a fixed angle relative to the direction of the source of a stimulus, but they do not necessarily involve locomotion. The dorsal (ventral) light reaction, displayed by many insects, is an example in which locomotion is not usually involved. Yet it is important as a means of maintaining primary orientation. The dorsal (ventral) light reaction is one means by which free-swimming aquatic insects maintain their normal position; back swimmers, *Notonecta*, depend upon a ventral light reaction for maintenance of their seemingly upside-down position, whereas the position of the water boatmen, *Corixa*, is maintained by a dorsal light reaction.

The most common form of transverse orientation involving locomotion is the **light compass reaction**. Since the locomotion is frequently oriented at a fixed angle, relative to a light source, it is an important aspect of navigation. Several simple experiments can be conducted to demonstrate this type of orientation. If a small black container is placed over an ant carrying food back to its nest (provided that it is not following a chemical trail) and is left there long enough for the position of the sun to change substantially, the ant, when released, will set out upon a new direction displaced by an angle equal to the angle of change in the sun's position (Figure 2–5). Similarly, if the image of the sun is blocked from the view of an insect navigating by a sun-compass reaction and a mirror is situated so that the insect sees a reflected image in a new position, it will change direction. The new track will be oriented at the same angle relative to the reflected image as the original track was to the real sun.

The sun, moon, and perhaps stars make particularly good navigational reference points because they are so distant that some insects are able to maintain a constant orientation angle and travel in a straight line for a long distance. If the light source is close, its angle of incidence on the

Figure 2–5. Course followed by an ant employing sun compass navigation. After spending 2 hours in a dark box at point *y*, the released ant changes the direction of its track by an angle (*x*) roughly equal to angle (*x'*) subtended by the arc through which the sun traveled during the period of captivity.

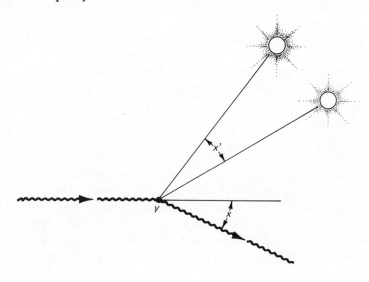

retina changes after the insect has traveled in a straight line for only a short distance, and the insect can only maintain its transverse orientation if it continually turns toward the source. This places the insect on a spiral course that ultimately ends at the light itself. It is in this manner that moths are thought to fly into the flame of a candle. Many more examples of responses to a variety of separate environmental stimuli could be presented, but they would serve little purpose. However, the principal characteristics of kineses, taxes, and transverse orientations are summarized for review in Table 2–1.

Under natural circumstances, insects are not exposed to a single dominant stimulus as they so often are in laboratory experiments conducted to gain an insight into a specific response. In nature, insects are exposed to a variety of stimuli to which they must respond in an integrated way, and most species display what appears to be a great deal of response compatibility. For example, foliage dwelling insects that are negatively geotactic are usually positively phototactic, whereas insects that live in the soil, are usually thigomtactic and display a negative phototaxis, a positive geotaxis, and a hygrokinesis.

COMPLEX INNATE BEHAVIOR

Many behaviorists recognize three kinds of external stimuli, each of which has a different effect. There are stimuli that **arouse** the individual, stimuli that **elicit** a response, and stimuli that tend to **orient** the individual during its response. In some instances, different intensities of the same stimulus may arouse, elicit a response, and orient an individual. For example, Shorey (1964) found that when a resting male cabbage looper detects a low concentration of the female's sex attractant, it begins to move its antennae and vibrate its wings (arousal); if a higher concentration of the sex attractant is detected, the male moth takes flight (elicitation), and then follows the concentration gradient to its source (orientation). More commonly, however, a series of different external stimuli are involved in the completion of such complex patterns of behavior.

Innate patterns of behavior displayed by all members of a species (or at least one sex of a species) are referred to as **fixed action patterns**. Although reflexes are in fact fixed action patterns, they have attracted less attention than more complex patterns. Some patterns of innate behavior are extremely elaborate; although their form is typical, the very first time they are performed their complexity gives the impression that the individual performing the pattern has had some prior experience. Clearly, these precise species-specific behavioral patterns have great adaptive significance in that the success and survival of a species depends, at least

Table 2–1. A Summary of the Characteristics of Oriented Behavior

Response	General Description	Type of Stimulus Required
Kineses	Undirected or random locomotor response, involving no orientation of the long axis of the body relative to the stimulus	A gradient of intensity
Orthokineses	The speed of locomotion, directly dependent on the intensity of the stimulus	A gradient of intensity
Klinokinesis	Frequency of turning, directly dependent on the intensity of the stimulus; can lead to aggregation	A gradient of intensity
Taxes	Directed reactions involving an orientation of the long axis of the body in line with the source of stimulation; toward stimulation—positive; away from stimulation—negative	A source of stimulation that generates a beam or a steep gradient
Klinotaxis	Orientation reactions involving regular alternating deviations of all or a part of the body across the field of stimulation	A beam of light or a steep gradient
Tropotaxis	Direct orientation to or away from stimulation by simultaneous comparison of intensities of stimulation to each side of the body	A beam of light or a steep gradient
Telotaxis	Direct orientation to a light source as if a goal; if two sources are present, orientation is directly to one source	A beam of light from a point source
Transverse Orientations	Orientation of the body with or without locomotion at a temporarily fixed angle relative to the source of stimulation	A point source of light, directed light or gravity
Dorsal (or ventral) light reaction	Orientation so that light is maintained perpendicular to both the long and transverse axes of the body	Directed light
Light-compass reaction	Locomotion in a direction temporarily fixed relative to the source of stimulation	Light from a small source

Source: After G. S. Fraenkel and D. L. Gunn, *The Orientation of Animals*, Dover Publications, Inc., New York, 1961.

initially, on their existence. In organisms such as insects, virtually all the vital behavioral functions that occur throughout the span of life are fixed action patterns.

Fixed action patterns normally cannot be performed unless the organism is in a state of readiness determined by both the internal and external environments. This prevents the organism from responding spontaneously to some stimulus when harm might result. When a state of readiness exists, however, the fixed action pattern is performed in a response to some specific form of stimulus called a **releaser**. For example, a newly emerged female with no ripe eggs in her ovaries would probably not be in a state of readiness for egg laying and, consequently, normal oviposition stimuli would not release oviposition behavior. Similarly, the stimulation of stretch receptors in the gut, caused by the presence of food, often inhibits the release of a normal feeding response; when the ingested food has been digested, the inhibition is lost and normal responses to food are displayed.

There is a range for most physical environmental parameters within which an insect is ready and able to initiate various patterns of behavior. When some condition (e.g., heat) falls below a lower threshold or rises above an upper thereshold, the insect will remain inactive even if the releaser stimulus is present. When the external environment is suitable and the insect is in a state of readiness, it may engage in activities that increase the likelihood of encountering a releaser stimulus. Most commonly, this so-called **appetitive behavior** consists of an increase of locomotion. Some bark beetles that have completed their migratory flight above the forest canopy will begin to fly about within the stand where the chances of coming in contact with chemical cues produced by others of their own kind are greatly increased.

Often, the inhibition or release of normal patterns of response are under the control of inherited, internal rhythms that induce behavioral periodicity. The diurnal rhythms displayed by many insects provide ample evidence of some internal "biological clock." For example, cockroaches are usually active during the hours of darkness and inactive during daylight hours. However, if the time of light and dark periods are transposed, the roaches may still be active at about the same time, even though it is light. The original rhythm may persist for several days before any adjustment to the new light-dark cycle occurs. The subject of activity rhythms will be considered in more detail in the next chapter.

There must be a large amount of coordination between both basic responses and various complex fixed-action patterns so that adaptive behavioral sequences are established that serve the organism's day to day needs. Often such sequences involve both inhibitory and releaser stimuli. Even in a rather simple process, such as feeding, a well-coordinated sequence of responses may be necessary. For example, blood-sucking flies will orient toward a host in response to the host's silhouette, odor, moisture, and temperature; as it approaches the host, the fly may extend its proboscis in response to an intensification of the same stimuli. A probing response may then be stimulated by odor, and the actual feeding by taste; feeding would stop when the fly was disturbed by a threat stimulus (perhaps some avoidance reaction on the part of the host) or when stretch receptors associated with a full gut inhibited the normal feeding response.

Other patterns, however, involve complex successions of responses, among which the nest-building and provisioning activities of solitary wasps and bees provide excellent examples. In some species the sequence can be interrupted only up to a certain point, whereas other species always seem to be able to resume an interrupted sequence where it was broken, and still others can switch from one sequence to another without apparent confusion. The potter wasp *Eumenes* will repair a hole made in its small earthen nest as long as it is under construction. But, once the wasp completes the nest and starts to provision it, she will not interrupt the latter part of the behavioral sequence even though the food she delivers may fall through the hole.

The solitary leaf-cutter bee *Megachile* also engages in an elaborate sequence of nest-building steps. The female must first locate a nest hole of appropriate size. She then leaves the nest site in search of a source of leaves and petals of a kind suitable for the construction of the individual larval cells. The cells are constructed of oval pieces of leaf or petal, cut to precision with the mandibles, and carried separately to the nest. When the first cell is complete, it is provisioned with 7 to 12 loads of pollen and topped with a load of nectar. An egg is then laid on the store of food, and the cell is capped with several discs of plant material. One such cell may take from 1 to 4 days to construct and provision, depending on the weather and the availability of resources. A series of such cells will be made sequentially, the number depending upon the depth of the nest tunnel. Each step in this process involves different stimuli, plus navigation to and from the nest, and each act must be done in the appropriate sequence.

The nest-building behavior of the solitary wasp *Ammophila campestris* is even more remarkable (Figure 2–6). The sequence begins with the construction of a vertical burrow in the ground that, upon completion, is sealed with several appropriately sized pebbles. The female wasp then leaves the nest site in search of a caterpillar to serve as food for her larva. When she finds a caterpillar, the wasp stings it in successive segments to paralyze it and carries the immobilized prey back to the nest. The small stones blocking the entrance are removed so that the caterpillar can be dragged into the burrow, after which a single egg is laid upon it. The wasp then leaves the burrow and temporarily seals it with the same pebbles selected earlier. The female then collects additional caterpillars, temporarily sealing the nest after each one has been added. Only after

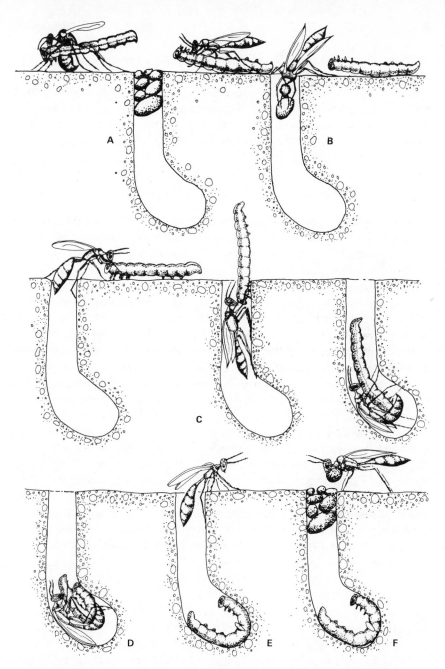

Figure 2–6. The successive acts in the provisioning of a previously constructed nest by the solitary wasp *Ammophila campestris*. After returning to the nest with a paralyzed caterpillar (*A*), the female removes the stones she has placed in the entrance (*B*) and drags the prey into the nest (*C*). The female then lays an egg on the caterpillar (*D*), climbs out of the nest (*E*), and replaces the stones (*F*). (Adapted from G. P. Baerends, *Jur. Tidjschr. Entomol.,* **84:** 68–75, 1941.)

the burrow is fully provisioned does the female engage in the final nest closing, which includes concealing the entrance by smoothing the dirt over it. What seems even more amazing is that *A. campestris* may have several nests in different stages of provisioning and can switch from one to another without any apparent confusion; in each case the sequence of steps dictated by the current status of each nest is adhered to rigidly.

LEARNED BEHAVIOR

Learning has been variously defined, but almost any definition that implies some beneficial adaptive change in behavior as a result of experience is acceptable. Learning implies some storage and retrieval of information (memory) that has a subsequent effect on the animal's behavior. Learning, therefore, improves the chances that an individual will be successful by eliminating unnecessary expenditures of energy, increasing the likelihood of obtaining a reward or reducing some form of punishment. Consequently, it is not possible to measure learned behavior directly but only to determine the extent to which the learning process has progressed. Among the insects, it is not a simple matter to determine what is innate behavior and what has been learned, because their complex patterns often appear to have a well-defined purpose. Some behavioral acts of insects even give the impression of insight; the use of a pebble by a female *Ammophila* to tamp and smooth the soil over her nest provides a good example.

Not all behaviorists are in agreement as to how different kinds of behavior should be organized, and entomologists are not in full agreement as to the kinds of learning displayed by insects. Alloway (1972), for example, identifies **conditioning, instrumental learning, shock avoidance learning,** and **olfactory conditioning** as four examples of learned behavior experimentally demonstrated among insects. However, other behaviorists claim that insects also display **habituation, latent learning, imprinting,** and even **insight learning**. In the following brief review I will follow the classification of Thorpe (1963) shown in Table 2–2.

HABITUATION

Habituation is recognized as the simplest form of learning. Unlike other forms of learned behavior, it involves the loss of a fixed action pattern rather than the development of some new response. Basically, an organism learns not to respond to a stimulus that has neither reward nor punishment associated with it. Insects are routinely exposed to many stimuli, the response to which would result in a wasteful expenditure of time and energy; learning not to respond would therefore be beneficial.

Table 2–2. Classification of Learned Behavior with Indication of Current Evidence of Learning in Insects

Type of Learned Behavior	Comments Regarding Insects
Habituation	Probably common among insects (e.g., reaction of mosquito larvae to shadows)
Associative learning Classical conditioning	Some experimental evidence of insect capability (e.g., proboscis extension by blowflies)
Instrumental learning	Some experimental evidence of insect capability (e.g., maze experiments with ants; feeding studies with honeybees; shock avoidance reaction of cockroach)
Latent learning	Some evidence from insects (e.g., orientation flights and homing among nest-building hymenopterans)
Insight learning	Circumstantial evidence only for insects; probably refined innate behavior
Imprinting	No good evidence of imprinting in insects

Source: After W. H. Thorpe, *Learning and Instinct in Animals,* 2nd ed., Methuen & Co. Ltd., London, 1963.

For example, mosquito larvae often spend considerable time at the surface of the water, periodically wiggling their way downward as an escape response to a shadow cast by an object overhead. However, they learn not to respond to the frequent passage of shadows that pose no threat, for example, those cast by overhanging foliage swaying in the breeze. Many such examples of habituation must occur throughout the Insecta.

ASSOCIATIVE LEARNING

Classical conditioning

Classical conditioning, also called conditioned reflex or respondent conditioning, was first described by Pavlov (1927) following his now famous study in which dogs were conditioned to salivate in response to the sound of a bell by repeatedly ringing the bell immediately before a food reward. In all examples of classical conditioning presented since, an unconditioned stimulus (UCS) that elicits an unconditioned response (UCR) is provided immediately following a conditioned stimulus (CS) to which the animal would normally show no response. However, when the UCS and CS are presented in sequence over and over again, the animal becomes conditioned to respond to the CS when it is presented alone.

Studies involving honeybees have long been cited as evidence for classical conditioning in insects. When offered a mixture of a sugar syrup reward (UCS) and the chemical coumarin (CS), the bees respond by extending their probosces. Studies have purported to show that when the usual UCS with withheld, the test bees would still display the reflex response to the coumarin. However, most insect behaviorists have considered the results of these studies to be inconclusive. Nelson (1971) provided more convincing evidence of classical conditioning in insects with his carefully controlled experiments involving proboscis extension by blowflies.

Instrumental learning

Instrumental learning, also known as instrumental conditioning and learning by trial and error, involves the modification of fixed action patterns through the application of **positive reinforcers** (rewards) and/or **negative reinforcers** (punishment). The most common form of instrumental learning involves the use of a maze. The basic technique is to attempt to improve a test animal's performance in the maze (increased response) through reinforcement with a reward (usually food); if the reward is not given, the response does not improve. Schneirla (1953) showed that it was possible to improve the ability of the ant *Formica pallidefulva* to pass through a maze (make fewer mistakes) by providing a food reward as a positive reinforcer. The learning curve generated in Schneirla's study is presented in Figure 2–7.

Shock avoidance learning and olfactory conditioning, identified as separate kinds of learned behavior by Alloway (1972) are considered by most behaviorists to be forms of instrumental learning. However, in some respects they bear more similarity to habituation than learning by the process of trial and error. Horridge (1962) provided the best evidence for insect shock avoidance learning. He was able to train cockroaches not to lower one leg beyond a certain level by attaching the leg to an electrical circuit that generated a shock (negative reinforcement) when it was lowered into a saline solution. The olfactory conditioning demonstrated in *Drosophila* also suggests learning. When larvae were reared in the presence of peppermint oil, normally an adult repellant, the resulting adults were attracted to it (Romoser, 1973).

The experiments conducted to demonstrate the color and chemical senses of bees have also provided insight into their learning ability. When a dish containing a 50 percent sugar solution is set among similar dishes containing distilled water on a checkerboard of colored squares, foraging bees will soon aggregate at the dish containing sugar. If that dish is subsequently switched with a water dish on a different colored square, the bees will continue to visit the dish on the colored square where they previously found the syrup. This association of a stimulus (color) with a reward (sugar syrup) clearly qualifies as learning.

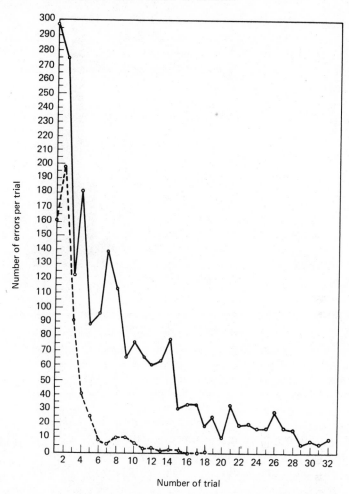

Figure 2–7. The learning curve for the ant *Formica pallide-fulva* in a six-point maze (solid line) compared to that for hooded rats (broken line). (Adapted from T. C. Schnierla, *in* K. D. Roeder, ed., *Insect Physiology*, John Wiley & Sons, Inc., New York, pp. 723–747.)

If a few bees are marked when they visit a feeding station placed on one square of a colored checkerboard, it is possible to construct a learning curve similar to that for the ants traversing a maze. The more experience each marked bee obtains, the fewer wrong landings it will make even when the position of the learned colored square is changed. Bees also seem to be capable of learning the association between the time of day and the availability of food. When presented with sugar syrup at a feeding station at a particular time for several consecutive days, they learn when to visit the station at the appropriate time. If food is subsequently made available all day, the bees will only visit the feeding station at the time food has been

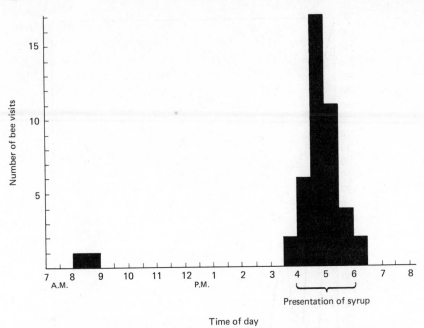

Figure 2–8. Graph showing the number of bees that visited a feeding place each half hour on a day when syrup was left out all day. Over the previous 3 weeks, syrup was put out only between 4 and 6 P.M. (Adapted from I. Beling, *Z. vergl. Physiol.*, **9:** 259–338, 1929.)

there previously (Figure 2–8) until they learn the new situation. Under natural conditions, this time memory is believed to be important in enabling the bees to visit certain flowers only during those hours when they make their pollen and nectar available (Wigglesworth, 1965).

LATENT LEARNING

Latent learning has been defined by Thorpe as ". . . the association of indifferent stimuli or situations without patent reward." Although this definition suggests a special variant of associative learning, the lack of an obvious reward or some immediate benefit creates an interesting distinction. Many insects, particularly the nest-building hymenopterans, make reconnaissance flights during which they display a capacity to learn the position of their nest relative to landmarks and celestial cues. Subsequently, these insects are able to locate their nests without difficulty following foraging trips at some distance (see Chapter 5).

INSIGHT LEARNING

Behaviorists recognize insight learning to be the most advanced form of behavior. The employment of insight, or solving problems by reasoning rather than trial and error, is a technique that humans take for granted. As a result, when we see an individual of another species solve a problem too rapidly to have conducted any trials, we conclude that insight was involved. Insects often display behavior that creates the impression that they are capable of insight. For example, a female *Ammophila* returning to her nest on foot, with a caterpillar slung beneath her, will almost instantly detour around obstacles placed in her path.

The use of a tool to perform some task is usually considered an example of insight learning, and there are at least two examples of tool using among the insects. As noted earlier, *Ammophila* females have been seen to use a small pebble held in the mandibles to smooth the ground over their nests, and the ant *Oecophylla smaragdina* holds one of its own larvae in its mandibles and uses the silk the larva produced to tie leaves together as a nest (Figure 2–9). However, if all or a reasonable percentage

Figure 2–9. Workers of the ant *Oecophylla smaragdina* drawing together leaves that are tied in place with silk produced by the larva held in the mandibles of the worker on the right. (Drawing by F. Nanninga, from *Insects of Australia*, Melbourne University Press, Melbourne, Australia, 1970. With permission of the Melbourne University Press.)

of a species engage in the same behavior, it is clearly a fixed action pattern even though it may appear to be insight.

There seems to be little doubt that the behavior of insects is dominated by inherited reflexes and response patterns and that innate behavior serves the insects well. There is also little doubt that some species of insects are capable of modifying their behavior on the basis of experience. Obviously, we have much yet to learn about the behavior of insects. With careful experimentation and sophisticated new techniques, the years ahead will almost certainly reveal some remarkable discoveries.

CHAPTER 3

Behavioral periodicity and clocks

Insects are the most successful group of primarily terrestrial animals. As such, they are subject to substantial diurnal and seasonal fluctuations in a variety of physical factors. In the long course of their evolution, insects have developed behavioral strategies that both exploit and lessen the impact of these fluctuations. Most insects, for example, display daily cycles of activity and quiescence. They may be active at night (**nocturnal**), during the day (**diurnal**), or during the transition periods of dawn and dusk (**crepuscular**). Many insects also display seasonal patterns of reproduction, growth, development, and dormancy that are of adaptive significance.

Behavioral periodicity serves as more than a means of avoiding periods of physical adversity and exploiting periods of physical favorability. As an inherited component of most living things, behavioral rhythmicity has become a significant part of coevolutionary relationships. For example, different plants bloom at different times of the year, and during the flowering period may only produce nectar and open at certain times of the day. Pollinating species of insects adapted to these plants must utilize a similar periodicity.

Many periodic phenomena are simply responses to periodic changes in the environment and are called **exogenous rhythms**. For example, some insects will remain inactive as long as the temperature is below some critical point. When the temperature rises above this inherited threshold, they become active regardless of the time of day. When the external stimulus is light intensity that changes regularly with the diurnal cycle of light and dark, the associated behavioral response can have a very regular periodicity. Exogenous rhythms obviously serve a multiplicity of functions but are subject to environmental variations that would not influence the behavior of organisms with a rhythmicity driven from within.

Internally driven, or **endogenous rhythms**, are hereditary endowments that manifest themselves even when external factors remain unchanged. Endogenous cycles with a periodicity close to 24 hours are called **circadian rhythms** (*circa,* about; *diem,* day). This is the most common form of periodicity observed among insects, but rhythms with annual (**circannual**), lunar (**circalunar**), and even tidal (**circatidal**) (Evans, 1976) periodicities are known. Sometimes the rhythmicity characterizes an event that occurs only once in the life of an individual, as in the case of the eclosion of *Drosophila* pupae at dawn. These timed, single events indicate clearly that endogenous rhythms are neither learned nor imposed by the environment, and the apparent innate ability of organisms to measure the passage of time has led to the use of the term **biological clock**. *Insect Clocks* is the title of an extensive recent review by Saunders (1976).

Behavioral periodicity can be observed in individuals and populations, and in most cases involves a combination of endogenous and exogenous components; the intrinsic clock mechanism provides a temporal organization that is then modulated by environmental periodicity. At the level of the individual, rhythms of general activity, feeding, mating, and oviposition are commonplace.

The rhythmicity of general locomotor activity of a number of insects in relation to light/dark cycles has been observed. Cockroaches and house crickets, for example, are principally nocturnal and begin their active period soon after dusk. However, behavioral periodicity in a light/dark cycle does not necessarily mean that the rhythm is endogenous. This must be determined experimentally by transferring the organism into continuous light or continuous darkness while other variables are held constant. If the activity remains rhythmic, with a periodicity close to 24 hours, the existence of an endogenous clock mechanism not controlled exogenously by environmental stimuli is indicated, but not proved. Under constant light or darkness and constant temperature, cockroaches maintained a locomotor rhythmicity that varied between 23 and 25 hours for several weeks (Roberts, 1960). Roberts also showed that the period of locomotor rhythm was shortened by an increase in temperature, but so little as to suggest virtual temperature independence.

Some insects, particularly crepuscular species, display bimodal activity patterns with peaks at dawn and dusk. When subjected to constant dark, some species will be active only once during the diel suggesting one endogenous period and one exogenous period. Others, however, display a bimodal periodicity in constant darkness.

An endogenous regulator of activity would have little functional value, unless it timed the activity to coincide with a period of over-all environmental favorability. Experiments have shown, however, that when endogenous oscillators are subjected to an environmental light/dark or temperature cycle, the endogenous periodicity becomes the same as that of the environmental cycle or is entrained by it provided that it is within the oscillator's range of adaptability. For example, when *Aedes aegypti* are held under constant light, they oviposit arrhythmically; but, when provided with a time marker (*Zeitgeber*) such as a dark period, they oviposit synchronously at daily intervals thereafter (Corbet, 1966).

Although the timing of events can be important to individuals, it is even more important to populations of mixed age groups as a means of synchronizing certain activities. This synchronization is achieved in part by once-in-a-lifetime occurrences controlled by an on-going circadian rhythm. The emergence of adult insects from their pupae is probably the best known of these phenomena, but a variety of others have been observed (Remmert, 1962). Clearly, the synchronization of certain behavioral events in mixed age populations has a distinct selective advantage. For example, egg hatching at dawn could coincide with humidity and temperature conditions less likely to be injurious to the newly emerged larvae. The synchronous early morning hatching of gypsy moth eggs, followed by the ascent of the young larvae to the upper branches of their host tree from which they drop on a silken thread, appears to be an adaptation related to their passive dispersal by wind. The synchronized emergence of new adults, resulting from a circadian pattern of pupal eclosion, would enhance the mating process, on one hand, but might result in extensive population inbreeding, on the other.

For any single species there can be several activities that display a circadian rhythm. The pink bollworm, *Pectinophora gossypiella,* for example, displays rhythms of egg hatching, pupal eclosion, and oviposition (Pittendrigh and Minis, 1971). When entrained to the same light cycle, all three show a different phase relationship to the *Zeitgeber*. These observations indicate the presence of a circadian system composed of a series of clocks, rather than a single "master clock."

The daily pattern of light and darkness serves not only as the most important *Zeitgeber* of circadian rhythms but as the most reliable indicator of the seasons and regulator of seasonal activity. The length of the light period is always the same on any given date at any given place on the earth (Figure 3–1). The ability to measure daylength, therefore, provides an organism with an annual clock that can be used to time accurately important events in its life. The most frequently observed response to photoperiod is the seasonal occurrence of dormancy in the life cycle. Most insects are active during the summer, and therefore develop and reproduce under a regime of long days, but become dormant under short days. This type of response to photoperiod is displayed clearly by species that have several generations per year, the last of which enters a facultative winter diapause. The stimulus for diapause is often a daylength of some critical value. In most species a daylength difference of 1 hour or less will determine whether development proceeds or is temporarily suspended.

The regularity of the photoperiod makes it a particularly useful advance indicator of seasonal adversity, but the daylength used to time diapause

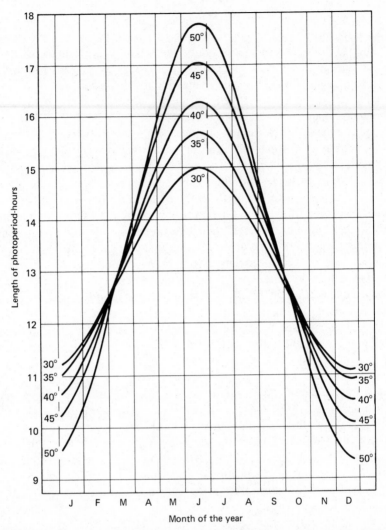

Figure 3–1. The relationship between photoperiod, date, and latitude.

varies from species to species according to their life cycle and the geographic location of their habitat. We usually think of diapause as a specialized overwintering state induced in many species by the photoperiod that characterizes the insect's summer period of development. However, in hot, dry areas many insects are active during the autumn, winter, or spring and pass the hot, dry period in a summer diapause induced by the short photoperiod that prevailed during their development.

As all daylengths, except those at the solstices, occur twice each year, investigators have sought evidence that insects respond to the direction of the change in photoperiod rather than to a stationary period above or below some critical value as used in most laboratory experiments. However, the life cycles of most species rule out the need for information regarding the direction of photoperiod change, since their responsive stage is only present during one of the periods when the appropriate photoperiod occurs. Nonetheless, there is evidence that some species do respond to the change in daylength that occurs between different stages of development. For example, when the immatures of the red locust, *Nomadacris septemfasciata*, are exposed to daylengths over 13 hours and the adults to daylengths of less than 12 hours, they enter a reproductive diapause that prohibits reproduction during the dry season (Norris, 1965). In the bollworm, *Heliothis zea*, diapause is induced when the larvae experience shorter days than the adults and eggs (Adkisson and Roach, 1971).

Long-range timers or endogenous rhythms with a period close to a year (circannual rhythms) have attracted more attention as a mechanism that regulates the reproduction and migration of longer-lived animals, like birds and mammals, rather than of insects. But as Saunders (1976) pointed out, the report by Blake (1958, 1959) of a circannual rhythm that governs the seasonal cycle of the carpet beetle, *Anthrenus verbasci*, represents the pioneer work in this field.

The life cycle of *A. verbasci* takes about 2 years for completion. Normally, 2 winters are passed in larval diapause, with the adults emerging the second spring. Blake (1958) demonstrated the endogenous nature of this rhythm by rearing the beetle in the laboratory under conditions of constant temperature, humidity, and darkness. When allowed to free run under these constant conditions, and therefore without a *Zeitgeber,* a natural periodicity of about 41 weeks was revealed. However, when the larvae experienced naturally decreasing daylengths, the cycle became entrained to the environmental year and pupal development was delayed to a time that would coincide with the spring. The 2-year life cycle could be shortened to 1 year under constant temperatures of 22.5 and 25° C, but the time of pupation was still dictated by the circannual rhythm to occur during an "allowable" period. At lower temperatures, an increasing number of larvae were unable to pupate during this first period; however, instead of their pupation being delayed only a few weeks, it was delayed until the pupation period in the second year. This form of endogenous temporal organization of an event is referred to as **gating** (Figure 3–2).

The endogenous timers discussed thus far control basic activity and developmental patterns that, in turn, influence responses per se and certainly must be understood by anyone interested in all forms of behavior. There are two additional types of time controllers that underlie behavior patterns with specific functional significance. The "time memory" of the honeybee, *Apis mellifera,* and the time-compensated sun-compass orientation of a variety of insects both appear to be under endogenous control.

Beling (1929) trained bees by providing sugar syrup to them at a

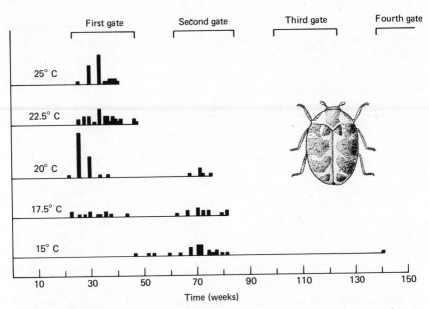

Figure 3–2. The circannual rhythm of pupation in *Anthrenus verbasci* showing the frequency of pupation times when larval development has occurred in constant conditions of temperature, humidity, and darkness. A black square represents the time of pupation, to the nearest week, of an individual. Note that the larvae at higher temperature are able to utilize the first gate; at lower temperature an increasing proportion of them are required to wait until the next. (From G. M. Blake, *Nature*, **183:** 126–127, 1959. Published by the courtesy of the Pest Infestation Control Laboratory, Ministry of Agriculture, Fisheries, and Food, Slough, England. Crown Copyright is reserved by the Controller, Her Britannic Majesty's Stationery Office.)

specific time of the day and marked some individuals while at the feeding station. Subsequent to the training period, no syrup was put out, but the time of arrival of marked bees was recorded. With this and later experiments, Beling demonstrated that the bees returned at the same time each day and that they could be trained to come at two different times of the day provided the periods were separated by at least 2 hours. The experiments of Beling, however, did not rule out the possibility that the bees were responding to exogenous time clues such as the position of the sun.

It was not until the 1950s that the ultimate proof of an endogenous clock controlling the time memory of the honeybee was forthcoming. Renner (1957; 1959) conducted experiments under both controlled and outdoor conditions that involved the westward relocation of trained bees. In a controlled experiment, bees transported overnight from Paris to New York (a real time difference of 5 hours) visited the New York feeding station exactly 24 hours after the previous feeding experience in Paris.

In an experiment conducted outdoors, bees transported overnight from Long Island, New York, to Davis, California (a real time difference of 3.25 hours), foraged at Davis 24 hours after the previous Long Island feeding; however, after only 3 days, there were indications of entrainment to the California light cycle.

The adaptive significance of the capability of keeping track of the temporal distributions of resources is obvious. Where this ability has been demonstrated, in the case of bees, it enables them to visit forage areas when pollen and nectar are readily available and even "remember" the time relationship during short periods when foraging is curtailed by poor weather. But as Saunders (1976) stated, "The fact that the rhythm is fairly easily extinguished without positive reinforcement, however, is also of biological importance because there is an ever-changing array of nectar sources, and there is little selective advantage in continuing to arrive at flowers long past their best."

Transverse orientations are common among insects, and one form, sun-compass orientation, is used by a variety of insects to traverse open space where landmarks are scarce. Insects using sun-compass orientation throughout much of the day must have a means of compensating for changes in the sun's azimuth. Early experimental results revealed that foraging ants held captive in a dark box for several hours maintain the

Figure 3–3. Time-compensated sun orientation in the honeybee. (A) A beehive H was placed in an unknown region, and a group of bees was fed *in the afternoon* on a feeding table F, 180 m NW. (B) During the night the hive was translocated to another area and *in the morning* the bees had to choose one of four feeding tables 180 m NE, NW, SE, or SW, of the hive. The new landscape did not offer any familiar landmarks; the sun stood at another angle relative to the training line as in the previous afternoon. Nevertheless, most bees (encircled numbers) came to the NW; that is, the bees had calculated the sun's movement. (After M. Lindauer, *Cold Spring Harbor Symp. Quant. Biol.,* **25:** 371–377, 1960.)

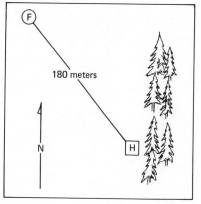

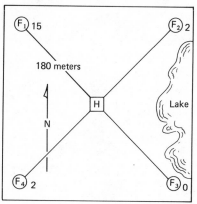

A   B

precapture angle of orientation relative to the new position of the sun when released. These results were counter to the possibility of time compensation. However, more recent investigations by Jander (1957) found that the black ant, *Lasius niger*, maintained its original compass bearing after release from a dark box. Similar results were obtained by von Frisch (1967) working with the red ant, *Formica rufa*, during the summer, but in spring experiments red ants seemed unable to compensate for the sun's movement.

Von Frisch (1950) also demonstrated time-compensated sun orientation for the honeybee. Bees were trained to forage at a feeding station west of their hive late in the day. The hive was then moved to an unfamiliar location during the night, and, when the bees were released in the morning, they foraged away from the sun (to the west) instead of toward it. Figure 3–3 shows the design and results of a more elaborate experiment with honeybees conducted by von Frisch and Lindauer (1954). The work with bees, ants, and a few other insects indicates the existence of some internal timing mechanism that enables them to compensate for the sun's movement. However, more work is needed on the light-compass orientation of insects at different latitudes and seasons of the year because of the differences in the apparent rate of the sun's movement.

PART II

FUNCTIONAL ASPECTS
OF BEHAVIOR

As stated earlier, patterns of behavior have evolved along with related anatomical features and physiological processes such that vital biological functions can be performed with a satisfactory degree of proficiency. Although we can identify these biological functions and the related patterns of behavior, they are closely intertwined and often interdependent. In some species the life cycle appears to be partitioned into periods during which one functional aspect of behavior seems to be predominant—a period of feeding is followed by migration, which is in turn followed by a reproductive period. In other species the behavior seems to contain a series of feedback loops that result in no clear or predictable sequence of behavioral events. Clearly, an experimental analysis of the over-all pattern of behavior would be exceedingly complex, and most behaviorists tend to investigate some aspect of behavior that has some basic biological interest or practical significance.

Markl and Lindauer (1965) suggested that the biological functions of behavior could be divided between the need of individuals to develop and sustain themselves, on the one hand, and interact with other individuals in order to contribute to the persistence of the species, on the other. Such a division might have some utility, but not all aspects of behavior are assignable under this particular dichotomy. Obviously, an individual insect must locate and select an appropriate type of food not only to sustain itself, but often to provide the nutritional requirements for the production of the eggs that become the next generation.

When viewed as an ongoing aspect of a species' biology, behavioral patterns and the functions they serve form a circular sequence rather than a progression with an obvious start and finish. When presenting an overview of behavior, the starting place at which one breaks into the sequence, is a matter of preference. After wrestling with the problem for some time, I selected a starting place that seemed to me to provide a logical flow with a minimum of backtracking.

I have chosen to begin with a discussion of behavior that leads to displacement and then consider orientation, navigation, and homing, because the ability of highly motile organisms like insects to move about in an organized manner must have great functional significance both locally and during their migration. It would then seem logical to consider communication in that the spatial separation of individuals, which often results from displacement, must be counteracted by different forms of aggregation behavior. The new aggregations thus formed may create a need for territorial behavior, as well as provide an opportunity for courtship, mating, and reproductive interactions among the constituent individuals. After these topics have been discussed, I will progress to a consideration of food selection and self-defense as two vital functions of the behavior of the new generation. Finally, I will discuss parental care and nest construction, which, along with various kinds of social interactions, seem to me to be special strategies that have evolved as solutions to problems associated with those functions already mentioned.

CHAPTER 4

Displacement

The high level of motility that characterizes so many insects has long attracted the attention of behaviorists both as a source of fascination and as a subject of great practical importance. In spite of the widespread attention that insect flight has attracted, there is much yet to be learned about behavior during flight and to the importance of flight as a cause of qualitative and quantitative changes in populations. One hardly needs to scratch the surface of the literature on insect dispersal and migration to realize that accepted generalities have been elusive. Even a common understanding and acceptance of the terminology has failed to materialize, particularly in the case of migration.

Any change of location can be called displacement. It may come about as a result of the accidental passive transport of insects by wind, water, phoresy (in association with another organism), active locomotion, or by some combination of these. In spite of the frequency with which passive transport, particularly by the wind, becomes involved in the movement of insects, there is relatively little displacement that is completely accidental and, therefore, detrimental at the population or species level. Small insects are not blown at random and do not end up in inhospitable environments as frequently as one might expect. Obviously, this was a danger that existed from the very origin of flight, and it is highly likely that species that did not evolve behavioral mechanisms to reduce in-flight accidents did not survive. On the other hand, the wind provides an inexhaustible source of external energy that insects could and have exploited with the evolution of appropriate patterns of behavior.

The movements of insects can best be separated into those often referred to as trivial and those that are truly migratory. Trivial movements tend to be local and lacking in directionality of a type that leads to predictable displacement. The fluttering of a butterfly from plant to plant in a meadow and the intermittent pausing to feed on the nectar of flowers or to lay eggs on suitable host plants is a typical example of trivial movement. The insect tends to change direction frequently and traverse territory it has probably traversed before, rather than fly in a more or less straight line over new territory. We cannot describe trivial movements as random because they may involve a variety of active responses to various stimuli. It is during this kind of activity that insects often locate food, mates, or oviposition sites. The initial phase of these selective behavior patterns often involves the detection of host odors or pheromones that are unevenly distributed as odor plumes. The trivial movements result in the interception of these cues, which may then lead to some oriented movements such as a positive chemotaxis or klinokinesis. The degree of displacement that results from trivial movements may be either quite substantial or rather small, as illustrated in Figure 4–1. Usually, insects involved in trivial flights remain within the boundary layer (the relatively thin layer of the atmosphere immediately above the substrate, where friction retards air movement and creates turbulence), and within a rather local area. Trivial flight is also characterized by numerous pauses during which insects feed, lay eggs, engage in mating behavior, or simply rest. Flight is, of course, not a necessary component of trivial movement, and all trivial movements do not result in long-distance displacement. Many ground-crawling and soil-inhabiting insects substantially change their location during their daily search for food and mates although the degree of displacement may be restricted to within a relatively small home range or territory. In my view, the back-and-forth foraging of insects like wasps,

Figure 4–1. Diagram of a hypothetical track of an adult insect engaged in trivial flight. Such flights are usually interrupted by numerous brief stops (dots). The ultimate displacement between the starting point and termination point (broken line) is usually considerably less than the total distance traversed (solid line).

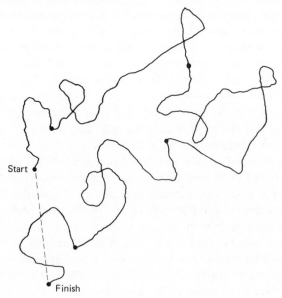

ants, and bees is a specialized type of trivial movement that results in only temporary displacement, since the foragers ultimately return to the nest, which remains in one place.

Migration, on the other hand, involves behavioral patterns that lead to a departure from one habitat and movement to another. In the case of aphids, it may result in movement from one host plant to another quite nearby; for other species, it may result in travel over hundreds of miles. Frequently, migration involves a behavioral pattern that leads to an escape from the boundary layer into higher elevations where the air moves horizontally at speeds that often exceed the intrinsic flight velocity of the species. Migratory movements are also more sustained than trivial movements and, in most cases, are characterized by a general straightening out of the track (Kennedy, 1975). Insects that migrate largely under their own power appear to be motivated by an inner drive that results in the suppression of their responsiveness to appetitive stimuli, such as food or mate odors, which would normally distract them during trivial flight.

Most of the confusion that has occurred in the area of displacement behavior involves the criteria for the definition of migration. Some early workers observed and recorded the mass flights of insects that often moved in a common direction with apparent purpose, or the sudden arrival of populations of insects in a new area. These investigators proposed definitions of migration clearly based on their limited experience. Some workers went as far as to suggest that true migrations are characterized by a two-way flight by the same individuals—clearly a carry-over from a basic understanding of bird behavior; one-way travel was described as dispersal. Many investigators also felt that migration resulted from a sudden deterioration of a species' habitat, which caused a mixed-age population to leave in search of more favorable conditions, and that the population was in full control of the direction of its travel. Small insects such as aphids, which commonly move short distances within the boundary layer, and the apparent drifters that comprised the aerial plankton sampled by men like Hardy and Milne (1938) and Glick (1939) were not considered to be true migrants. In fact, very few species actually fulfill the early definitions of migrants.

In the case of short-lived species such as most insects, there is no biological rationale for limiting migration to a back-and-forth pattern of movement by the same individuals. Many do not live long enough to travel in both directions. However, the fact that some insects do move back and forth between different parts of their habitat has confused the issue, to say the least.

Some of the best examples of regular back-and-forth travel within a local area have resulted from studies of a group of scarab beetles known as cockchafers or leafchafers. The adults of a number of species emerge from the soil of pasture areas where pupation occurred and remain in the grass until mating has taken place. The inseminated females then fly to nearby woodlands, where they feed on the leaves of deciduous trees and develop their eggs. Subsequently, they fly back to the pasture land where they oviposit in the soil. The result is the repopulation of previously occupied breeding grounds and little redistribution of the population to new areas. Although a rather special situation, this behavior does constitute migration of Johnson's type II discussed a little later. Some workers have extended this interpretation of migratory behavior to other cases of regular back-and-forth movement within a habitat. Certainly, such movements are behaviorally different from the trivial flights of the butterfly in the meadow described earlier, in that the flight path may be straight and the displacement predictable. However, in many cases the travel is accompanied by responses to feeding and reproductive stimuli normally ignored by migrants and consequently results in the fulfillment of basic needs normally a function of trivial flight.

Since most species sometime in their life cycles engage in displacement activity beyond that associated with trivial movement, we obviously need a biologically sound concept of migration. The key to understanding this aspect of behavior, therefore, seems not to be whether the population traverses a great distance, not whether the travelers orient themselves as though they have some destination in mind, not whether the same individuals will make a return trip, but that the species periodically engages in travel as an integral part of its life cycle and behavior. This is indeed the concept of migration presented by Southwood (1962), Johnson (1969), Kennedy (1975), and others. Kennedy (1975) has made one of the strongest recent attempts to clarify the situation by suggesting that most insect species must travel from one part of their range to another as a regular and vital part of their biology for purposes of resource utilization and gene pool mixing. Furthermore, the advantages of such travel have led to the evolution of specific behavioral adaptations that accommodate it.

CAUSES OF MIGRATION

The fact that insects are often observed to engage in a mass exodus from a breeding site, coincidentally with crowding or some deterioration in the quality or quantity of their food supply, has led to the conclusion that migration is simply an immediate behavioral response to the environmental conditions that exist. However, many species of insects seem to migrate as a matter of course even though they are not crowded and their habitat appears to be suitable for continued utilization. This occurs because many species actually leave their breeding place well in advance of its deterioration as an adaptive response to some early warning signals. The suggestion that migration is a sudden behavioral response to adversity is also weakened by the fact that migrants often continue to travel long after they have left the unfavorable area without testing the favorability of areas along the way. Still other species, which have several generations per year, migrate during each generation, not only the generation subjected to environmental deterioration.

Southwood (1962) concluded that migratory behavior is particularly important for species that occupy temporary habitats such as temporary pools or scattered hosts, or habitats that become periodically unsuitable, as with the seasonal drying up of vegetation. Occupants of such habitats must be adapted to migrate at more or less regular intervals in order to survive the year-to-year variability inherent in heterogeneous environments. For these species the migratory behavior, therefore, is best set in motion by environmental stimuli that precede the impending adversity. Photoperiod would seem to be the most reliable stimulus and one deeply involved in a number of periodic behavioral activities (see Chapter 3). However, an increase in population density, changes in food quality or quantity, and changing climatic conditions can all be used as advance warning signals if an appropriate response system has been evolved. From a behaviorist's point of view this may pose a problem because, late in the summer, insect population growth, high mean daily temperatures, drought, declining photoperiod, and a deterioration of the vegetation may all go hand in hand.

From the evolutionary point of view, it would seem that a mechanism that integrates the environmental cues with the insect's developmental physiology would have the strongest selective advantage. Experimental evidence suggests that this is accomplished by way of the environmental effects of the endocrine system. For example, there is an established relationship between crowding and the size and activity of the prothoracic glands. The prothoracic glands of crowded locusts are small at emergence and disappear a few days after the final molt, whereas in isolated individuals these glands are large and persist until sexual maturation is complete. Crowded individuals engage in more active locomotion than isolated individuals, and, when the latter are injected with hemolymph from crowded locusts, they too become active (Haskell and Moorhouse, 1963). Other relationships exist between environmental stimuli and activity of the corpus allatum and the development of ovaries. Poor-quality food and declining photoperiod will both result in retardation of ovation development, a condition usually encountered among migrants. Thus there is a considerable degree of coordination between the development of the flight apparatus and the ovaries in relation to various environmental changes. A generalized diagram of the more important interrelationships is provided in Figure 4–2.

Because of the incompatability of a well-developed flight mechanism and fully developed ovaries, migration tends to occur when ovarian de-

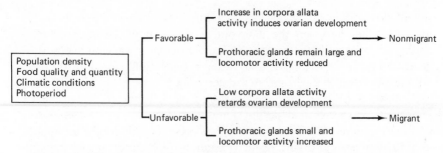

Figure 4–2. The interrelationships between environmental conditions, hormonal activity, and locomotion that may lead to the induction of either nonmigratory or migratory behavior.

velopment is attenuated. The balance between migratory flight and the reproductive system may vary according to the environmental conditions, as suggested previously, but in many species this balance is a regular part of development. In quite a large number of species, migratory flights occur soon after emergence. The extent and duration of such flights may vary greatly from individual to individual depending on their developmental histories, but they occur nonetheless. During these early flights, many insects are clearly nonresponsive to appetitive stimuli, as demonstrated by the fact that older individuals tend to dominate the catches of baited traps. The nonresponsiveness of young migratory individuals has been demonstrated in studies of the coddling moth (Geier, 1960), the screwworm (Crystal, 1964), the Douglas fir beetle (Atkins, 1966), and other insects.

Many observers have noted that not all individuals in a population migrate; some remain in the vicinity of their breeding site. When the electric buck moth, *Hemileuca electra* (Saturniidae), emerges in the fall, some of the females are so heavily laden with eggs that they are capable of only short, weak flights from plant to plant and, consequently, lay their eggs in the immediate vicinity of their pupation site. Others emerge, fly upward out of the boundary layer, and leave the area completely. In fact, these moths are recognized for their ability to fly strongly, well above the ground on windy days. Whether the behavioral differences displayed by individual moths is under genetic control or is the result of nutrition and developmental history remains unknown. In my work on the flight behavior of the bark beetle, *Dendroctonus pseudotsugae* (Scolytidae), I found a correlation between fat content and the tendency to display migratory behavior (Atkins, 1967). Hagen (1962) suggests that it is possible to identify migrant convergent lady beetles heading for hibernation sites by squeezing them to see if they contain much fat. Perhaps nutrition alone can cause the differences in behavior that would contribute to the exploitation of both local and distant habitats. However, we should not rule out the possibility and selective advantages of genetic differences.

Thus we see among the insects a confusing situation in which some groups such as the aphids, termites, and ants periodically produce winged obligatory migrants, whereas many other species produce only winged adults, some of which migrate and some of which do not. We see some migrations associated with crowding and environmental deterioration, and some that occur without environmental change. The actual cause of migratory behavior is still not clear, but we have now come to realize that there is an endogenous motivation genetically programmed into the behavioral make-up of most, perhaps all, insect species, and that migration is clearly an ecologically and genetically beneficial form of adaptive behavior molded by the process of natural selection. As Johnson (1963b) stated, "To the universal cycle of birth, reproduction and death must be added the process of migration . . . "

CLASSES OF MIGRATION

There is great variability in insect migratory activity. It ranges both behaviorally and ecologically from something not much different from trivial movement in terms of distance traveled to travel over very long distances. Johnson (1969) erected the following three classes of migration which provide considerable clarification:

Class I.   Species with a life span limited to a single season leave their breeding site, disperse to new areas to reproduce, and die soon thereafter. The migrants make only an outward journey, but their behavior is characterized by an impression of moving toward a goal.

Class II.   Species with relatively short-lived adults that leave their breeding site and travel to a new location where they feed and their ovaries develop. After maturation, the females return to their old breeding site or a new one where they oviposit.

Class III. Relatively long-lived individuals leave their breeding site, travel to a winter or summer resting site where they pass through a reproductive diapause. During the following season the same individuals return to their original breeding area, where the females oviposit.

Class I migrations may be extremely variable in terms of both distance traveled and duration of travel but usually occur soon after emergence and before the gonads have matured. Johnson recognized five subtypes of this class of migration, but a pair of examples should give an adequate impression of the variation that exists. A rather simple illustration is provided by the flights of ants and termites, in which winged individuals that are produced seasonally become the founders of new colonies. The winged males and females leave the nest soon after emergence and fly weakly with the wind. As the wind direction changes, the winged reproductives become displaced variously within and beyond the area where breeding is possible. Relatively few new nests need to be established to keep the area adequately populated, so the large numerical losses sustained during migration can be tolerated.

Although seemingly different and certainly of a greater magnitude, the migration of the desert locust, *Schistocerca gregaria* (Acrididae) is also an example of class I migration. In tropical and subtropical areas, the desert locust must move from breeding areas that are deteriorating because of drought and resource depletion to new areas that are receiving rain and have green vegetation. These breeding areas are often widely separated and tend to be in different locations because of spatial variations in the pattern of seasonal rainfall.

The discovery that the seemingly powerfully oriented flight of the desert locust across Africa was simply adaptive travel with the wind helped immeasurably to change our view of migration. Migratory locusts are not preprogrammed to orient toward some distant goal as once thought; instead they are adapted to take advantage of the seasonal wind patterns. The analysis of films of migrating swarms of locusts revealed that the individual migrants were not necessarily oriented to their direction of travel (track). Within a swarm there are many groups of similarly oriented individuals, but throughout the swarm these groups fly in different directions. The different orientation of these groups maintains the cohesiveness of the swarm, which as a whole moves with the wind. The swarm is thereby gradually displaced downwind to zones where tropical air masses converge and generate the rainfall that stimulates plant growth in the new breeding place.

The pattern of class II migrations is illustrated by many insects that display a high degree of larval-adult divergence in terms of habitat or food preference. Female mosquitoes, for example, must often leave the area of larval habitat from which they emerged so as to obtain a blood meal necessary for egg development. The larvae are filter-feeders in aquatic habitats, whereas the females often search for vertebrate hosts some distance from the water. After feeding and oogenesis, they return to aquatic habitats to oviposit. In the chafers mentioned earlier, the larvae feed on organic matter in the soil of pastures, whereas the adults feed on the leaves of trees to which they migrate. After maturation of their ovaries they return to the pastures to oviposit (Schneider, 1962).

Class III migrations vary considerably in detail from species to species.

At one extreme, young adults emerge from their breeding site and fly in all directions throughout their habitat, sometimes aided by local winds, coming to rest in a place suitable for spending a dormant period. Following dormancy, the same individuals migrate back to available breeding sites where oviposition takes place. The ambrosia beetle, *Trypodendron lineatum,* for example, emerges from brood sites in logging slash and flies into forest margins in the same general area. After passing a winter diapause in the litter of the forest floor, the beetles emerge and fly out into adjacent open areas, where they locate new brood logs. On the other hand, some lady beetles, one of which will be discussed in detail later, migrate long distances between breeding areas and hibernation sites.

Some of the longest and most predictable migratory flights known among the insects are of the class III variety. The monarch butterfly, *Danaus plexippus,* in North America migrates hundreds of miles to a rather well-defined overwintering area, from which it returns to breeding areas the following spring. The vast distance covered by monarch populations has made it difficult to determine exactly how far individuals fly during each of the phases of the migration. The breeding range of these butterflies is extensive north to south, so there is considerable variation in the latitude at which they develop. The adults that result from the last generation in Canada migrate to regularly used overwintering sites in California, Mexico, and Florida. The butterflies are active on warm days during the winter but engage only in trivial flights in the immediate vicinity of their overwintering site. In the spring, the overwintered adults fly northward to new breeding areas, but much less is known about the nature of the return flight. Some are believed to terminate their migration early and establish populations at intermediate locations, but others are thought to return directly to the northernmost breeding areas (Urquhart, 1960).

THE ADAPTIVE NATURE OF MIGRATORY BEHAVIOR

The energy required to sustain flight during an extended period of migration is substantial, and, although many insects have large energy stores at the beginning of their migratory flight, there is clearly an advantage to harnessing the energy of the wind. Lipids are the fuel with the best weight-to-energy ratio and are used by many sustained fliers such as locusts, butterflies, and beetles, although some flies with substantial flight ranges use carbohydrate as their fuel. Calculations suggest that the initial fuel supply available to migrants provides some of them with a great intrinsic flight capacity, and this has been borne out by flight duration tests made with insects tethered on flight mills. I have often obtained 8 hours of continuous flight from tethered Douglas fir beetles, *Dendroctonus pseudotsugae* (Atkins, 1961). We know very little about how much of their fuel supply migrating insects replenish en route, but the suppression of feeding stimuli, common at least at the start of migration, and the carbohydrate nature of most available foods suggest it may be very little. Nonetheless, some migrants such as the monarch butterfly are known to fly distances that exceed their intrinsic range as calculated from their fuel capacity and flight speed, and observations do suggest that the inexhaustible supply of external energy provided by the wind is exploited by many species.

The general pattern of atmospheric circulation is governed by the interaction of large pressure cells variously positioned according to the seasons. Near the ground these regional circulation patterns are modified by local climatic and topographic effects. The frequency with which local winds shift direction would make their general exploitation risky, in that it could lead to enormous numerical losses if insects were carried to inhospitable habitats. Insects, then, must be strong enough to combat the wind or have behavior that either restricts their flight to periods of relative calm or exploits winds that would normally carry them to favorable habitats.

Some insects, such as the painted lady, *Vanessa cardui* (Nymphalidae), are well known for their strong directional flights (Williams, 1970). In southern California during the early spring, mass flights of adult painted ladies can be seen heading northward from breeding sites in Mexico. Abobtt (1961) reported that these butterflies fly in a northerly direction, regardless of the direction of the wind. In 1973 I observed a large mass migration in the San Diego area that lasted several days. One day, the butterflies flew northward against a steady wind of about 15 miles per hour, and, on another day, the same flight path crossed an easterly wind of similar velocity.

Many insects that are weak fliers also have specific patterns of behavior that clearly enhance their opportunity to migrate with the help of the wind. Aphids are only capable of a flight speed of approximately 2 miles per hour, so are subject to displacement by even gentle winds once they let go of their substrate. Migratory individuals display a strong positive phototaxis. When they take off in response to the skylight, they fly with an excess of lift over horizontal flight. This carries them upward into the wind, which helps them to travel overland many miles. The aphids apparently beat their wings while being displaced downwind and thereby satisfy the tendency to sustain flight. This locomotor drive declines after an hour or two, and the aphids begin their descent. If an aphid lands on a suitable host plant, it will begin to feed and later reproduce; if it lands on an unsuitable plant, it will take off again. This behavior may be repeated several times over several days and can lead to travel over many miles. Many individual aphids are lost during these migrations (Johnson, 1960), but the synchronized exodus of winged females from their summer host plants assures the survival of the species on alternate hosts throughout the fall and winter.

The importance of adaptive flight behavior in relation to variable environmental conditions is also important and well illustrated by the migrations of the beet leafhopper, *Circulifer tenellus* (Cicadellidae), in the San Joaquin Valley of California, as summarized by Cook (1967). The leafhoppers overwinter and pass their first generation predominantly on wild vegetation at the southern end of the valley. In the spring, the prevailing northwesterly winds enter the valley in the area of San Francisco and are deflected southward through the valley; this is in the opposite direction of the spring leafhopper migrations. The leafhoppers are capable of flight speeds in the order of 2 miles per hour and tend to fly only when the prevailing winds have abated. This frequently occurs in the late afternoon when the warm air in the valley rises and is replaced by downslope winds from the surrounding hills. These winds create a northwesterly flow across the leafhoppers' breeding site that carries those already in flight well out into the valley where sugar beets are cultivated. As a result, the spring migration is against the prevailing wind. Similar combinations of leafhopper flight behavior and wind patterns at other times throughout the season allow the hoppers to move throughout the valley and ultimately end up at their hibernation sites in the southern hills in the autumn.

Studies of the pre- and postdiapause migrations (class III) of the convergent lady beetle, *Hippodamia convergens,* by Hagen (1962) revealed adaptive behavior patterns that enable these fairly weak flying beetles not only to travel substantial distances, but also to utilize habitats with rather specific locations. In years when aphid populations are high, large numbers of young adult lady beetles emerge from fields in the lowland valleys during May and June. The general reduction in the abundance of aphids because of prior feeding by the beetle larvae leads to the departure of the young adults. About this time, large numbers of lady beetles are often observed in the vicinity of aggregation sites in the mountains. The beetles apparently leave the fields by way of vertical take-off flights on warm, calm mornings. These vertical flights, assisted by conventional currents, continue upward to a temperature ceiling of 11 to 13° C, which curtails flight. The nonflying beetles are believed to fall into warmer air that permits a resumption of their upward flight. This alternating pattern of upward flight and falling produces oscillations of movement that may have an amplitude of up to 1000 feet (305 meters). During these oscillations the beetles are carried on horizontal, westerly winds toward the mountains, where they are deposited in the zone of intersection between the temperature flight ceiling and the ground level (Figure 4–3A).

Warm days during February and March in the Sierra Nevada are associated with a high pressure system over southern Idaho that produces

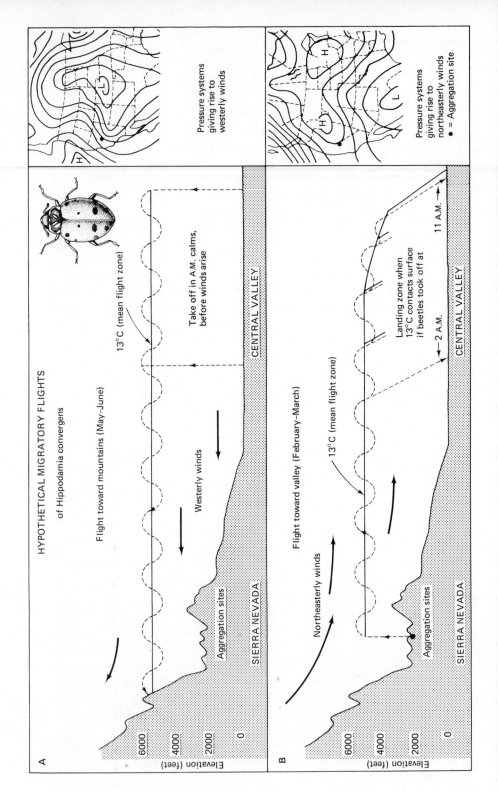

HYPOTHETICAL MIGRATORY FLIGHTS
of Hippodamia convergens

A

Flight toward mountains (May–June)

13°C (mean flight zone)

Take off in A.M. calms, before winds arise

Westerly winds

Aggregation sites

SIERRA NEVADA

CENTRAL VALLEY

Elevation (feet)
6000
4000
2000
0

Pressure systems giving rise to westerly winds

B

Flight toward valley (February–March)

13°C (mean flight zone)

Northeasterly winds

Landing zone when 13°C contacts surface if beetles took off at

11 A.M.

2 A.M.

Aggregation sites

SIERRA NEVADA

CENTRAL VALLEY

Elevation (feet)
6000
4000
2000
0

Pressure systems giving rise to northeasterly winds

● = Aggregation site

northeasterly winds aloft. The aggregation sites on the western side of the mountains are protected from these upper-level winds, but they experience convectional currents and warm upslope winds from the valley. When the temperature of the aggregation sites rises, the beetles break dormancy, fly upward, and eventually engage the winds aloft that carry them back to the valley. The beetles again oscillate up and down in the vicinity of the temperature flight ceiling as before but are forced to the ground as the temperature ceiling declines late in the day (Figure 4–3B).

The numerical losses that occur during insect migrations are clearly high, but the benefits that accrue must more than compensate for the in-flight population attrition. Insects that utilize scattered temporary habitats must produce enough progeny to ensure that a few survive the migration to new habitats. The fact that the habitats of such species are scattered unevenly throughout their range have been a strong selective force that favored a strategy in which migration and habitat location are the functions of the winged adult. Yet, this seems inefficient in respect to all the resources consumed during the development of the high percentage of adults that die during the movement between habitats. On the surface it would seem to be more efficient for dispersal losses to be absorbed early in the life history, but this occurs in only a relatively few species of insects. As one might expect, those that do migrate as juveniles are rather generalized feeders or users of a somewhat specific, but widely distributed, food source.

Those species that migrate as immatures have little capacity to navigate or terminate their travel. In order to migrate and maximize the success of establishment in a new habitat, appropriate adaptive behavior would seem to be a necessity. Studies of the behavior that leads to and terminates the passive dispersal of young caterpillars and first instar homopterans reveal that their behavior is truly adaptive and therefore migratory.

Migration by young larvae of the tussock moth family Liparidae is quite common. The first instar larvae of the gypsy moth, *Porthetria dispar,* display a rather precise pattern of behavior that leads to their transport on the wind. They display a periodicity of egg hatching, which occurs in the early morning, independent of the ambient temperature and humidity. Between 0800 and 1000 hours, and again, but to a lesser extent, in the afternoon, the small caterpillars ascend the trees and move out to the tips

---

Figure 4–3. (*At left.*) Suspected temperature-controlled flight behavior of *Hippodamia convergens* engaged in migratory flight. (A) The hypothetical pattern of migratory flights to overwintering sites in the Sierra Nevada in May and June. (B) The hypothetical return flight to the San Joaquin Valley of overwintered individuals in February or March. The simplified weather maps indicate the dominant pressure systems at the time of the respective migratory flights. (From K. S. Hagen, *Ann. Rev. Entomol.,* **7:** 289–326, 1962.)

Figure 4–4. Young larvae of the gypsy moth showing the long lateral setae believed to improve their passive transport by the wind. (Courtesy of the U.S. Department of Agriculture.)

of the branches. Between 1300 and 1500 hours, the larvae drop on strands of silk. At this time of the day both the horizontal and vertical air currents typically reach their greatest velocity. The suspended larvae swing back and forth until the wind is strong enough to break them loose, at which time they sail on the wind to new habitats. The long, lateral setae of the larvae (Figure 4–4) increase their buoyancy and appear to be an adaptation that aids passive transport.

OVER-ALL SIGNIFICANCE OF MIGRATION

Insects, like most organisms, derive both genetic and ecological benefits from being able to move from one place to another. Such behavior increases the mixing of the gene pool and hastens the spread of beneficial mutations. Ecologically, migration enables a species to vacate crowded areas or habitats, where the requisites for life are deteriorating, in favor of sparsely populated areas or habitats with an abundance of appropriate resources. As is usually the case with evolution, the persistence of a trait depends on whether or not it contributes to the survival of the species. The benefits of migration, therefore, must outweigh the associated physiological costs and numerical losses. Obviously, all insect species do not have the same need to migrate. Those with more or less permanent breeding habitats may be able to achieve adequate displacement by way of the general diffusion that accompanies trivial flight. At the other extreme, species that occupy very temporary or transitory habitats need to migrate regularly.

In order that the genetic and ecological benefits be obtained, substantial losses in the form of energy expended and of mortality must be borne. Apparently, these costs and benefits have been favorably balanced in the course of evolution. The evolution of adaptive behavior patterns has proved to be one way to maximize the migratory gains and to minimize the attendant losses. In fact, it would be surprising if many species of insects could survive long if they had to rely on some haphazard means of moving from one habitat to another. Even so, there would seem to be a limited number of behavioral strategies that can be exploited, even though there may be an infinite number of minor variations in the details of such behavior.

CHAPTER 5

Orientation, navigation, and homing

In addition to migratory movements, insects move around a great deal within what can be called their **home** or **action range**. For a given species the nature and purpose of these movements are much more variable than for migratory movements. As we noted in the last chapter, migrations are characterized by the straightening of the individual's track resulting from a suppression of its responsiveness to appetitive stimuli. Trivial movements, on the other hand, are characterized by frequent responses to appetitive stimuli. However, both trivial and migratory movements require some degree of orientation and some ability to navigate. The need to navigate and return to a specific location varies from species to species. Those which have adapted to migrate with prevailing winds, such as the desert locust or lady beetle discussed in Chapter 4, do not need to navigate in a strict sense although some specific orientation relative to the wind or other individuals may be important. Conversely, species that travel great distances independent of the wind, like some butterflies, must have some means of maintaining an appropriate course. Likewise, within their home range some species may need only to recognize the boundaries of their range, whereas others, particularly those that utilize nests, may need to navigate with considerable precision.

The fact that trivial movement involves the search for mates, oviposition sites, food, and so forth, suggests that stimuli associated with these basic requisites will often influence an individual's orientation. Within its home range then, we should expect to observe much behavior dominated by the orientation to specific stimuli such as host odors or the auditory call of a mate. These specific patterns of orientation and the kinds of stimuli that govern them will be discussed in Chapters 6, 7, and 8. In this chapter we will examine the means by which insects find their way within their home range, including their repeated return to nest sites and how they navigate during migration. What will be presented will, by necessity, be a mixture of experimental evidence and conjecture. Much of the work done on insect homing in the field has produced some solid results, but such is not the case for long-distance migrations.

NAVIGATIONAL CUES

Stimuli of the kinds normally associated with feeding and reproduction have a pronounced influence on orientation after the migratory drive has been satisfied and the individual is engaged in the search for food, mates, or reproductive sites. Appetitive stimuli would seem to be less important to navigation and homing. These latter processes would be served best by more reliable (in the sense of more accurately identifying location) reference points such as celestial cues (sun, moon, and sky polarization), landmarks (visually discernible fixed objects and patterns), electromagnetic fields, and in a few instances currents or flow patterns. For homing insects, chemical trail markers can also be important.

Celestial cues

A wide variety of insects are capable of maintaining the alignment of the long axis of their body at a fixed angle relative to a source of stimulation. These so-called transverse orientations may or may not involve movement. A commonly displayed example involving movement is called the light-compass reaction. Celestial cues such as the sun, moon, stars, and the polarization of the sky are particularly good reference points for a light-compass reaction because they are so far away that orientation relative to them can result in travel along a straight line for a long distance. The innate nature of transverse orientations is an explanation for the seemingly self-destructive behavior of insects flying into a candle flame or light fixture. When the artificial light source is close at hand, the insect can only travel in a straight line for a short distance before it must change direction to maintain its fixed angle to the source of stimulation. The insect travels along a logarithmic spiral that ultimately leads to the light source (Figure 5–1).

Light-compass orientation can be demonstrated by several simple experiments first conducted many years ago. While working with ants foraging in featureless deserts of Tunisia, Santschi (1911) shaded workers returning to their nest from the sun and provided them with a mirror image of the sun projected from the opposite side of their track; the ants turned 180° and headed away from their nest. Brun (1914) placed a worker of the ant *Lasius* in a small light-tight box on its homeward journey and held it captive for 1½ hours. When released, the ant traveled in a new direction, 23.5° from its original path. As the sun had traveled through an arc of 22.5° during the same time, it can be concluded that the ant was

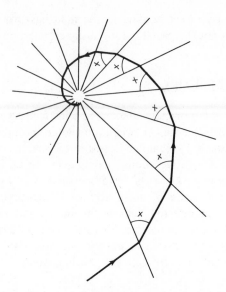

Figure 5–1. The spiral track followed by an insect orienting to a nearby point source of light. The "logarithmic spiral" results from the fact that the insect is actually employing light-compass orientation and attempts to maintain a line of travel at a fixed angle (x) to the light.

using a sun-compass orientation but was not capable of time compensation. The lack of a time-compensation mechanism would make celestial orientation ineffective for any insect that may forage away from its nest for some time and could be prevented from returning, as happens when bees are grounded by a drop in temperature associated with passing clouds. However, Jander (1957) demonstrated time-compensated orientation for experienced *Formica* workers, and a similar capability has been observed in the honeybee (von Frisch, 1950). The ability to navigate relative to the position of sun is, of course, best known in the honeybee, which has evolved an elaborate communication system based on the sun as the key reference point (see Chapter 6).

The phases of the moon and the ease with which it is obscured by clouds would seem to make the moon a less useful navigational cue for nocturnal insects than the sun is for diurnal species. Stars also would seem to be marginally useful reference points for insects, and I know of no evidence that they use them. Many night-flying insects do display a light-compass response, however, and a moon-compass orientation has been demonstrated in the ants *Monomorium* (Santschi, 1923) and *Formica* (Jander, 1957).

Huffman (unpublished) has worked extensively with the waterboatman, *Trichocorixa reticulata,* to determine the basis of its escape orientation away from the shoreline toward deeper water. One might assume that the escape could be achieved readily by following the slope of the bottom or by utilizing depth receptors, but both were ruled out experimentally. Huffman has shown that during the day the sun and to a lesser extent polarized skylight are the main orientation cues, whereas at night there is strong evidence of moon-compass orientation. Since shore birds that prey upon *Trichocorixa* are active on moonlit nights, the moon-compass orientation provides an escape mechanism not needed when the moon is absent. It is tempting to conclude that the light-compass orientation of many nocturnal species represents an evolutionary carry-over that may be beneficial when the moon is bright enough to provide a navigational reference point.

Some insects can also detect polarized light and therefore orient relative to the position of the sun even though they may not be able to see its actual disc. The light waves produced by the sun vibrate in all directions at right angles to their direction of travel, but, as they pass through the earth's atmosphere, the light is scattered and some planes of vibration are eliminated until each wave tends to vibrate in a single direction at each point in the sky. The resulting pattern of polarized light varies according to the relative position of the observer, the position of the sun, and the portion of the sky being viewed. The area of maximal polarization forms a band across the sky at approximately 90° to the sun's angle above the horizon and centered about the opposite point on the compass. For example, when the sun is 45° above the southeast horizon, the maximum plane of polarization would stretch across the northwest quadrant of the sky approximately 45° above the horizon. Consequently, the plane of maximum polarization shifts as the sun moves across the sky (Figure 5-2). Although the phenomenon of atmospheric light polarization has been known since before the turn of the century, it was not until the late 1940s that Karl von Frisch discovered that honeybees navigate by polarized light (von Frisch, 1949). Only very recently has the mechanism of polarized light perception in insects been determined (Wehner et al., 1975; Wehner, 1976).

The early work of von Frisch stimulated a number of insect behaviorists to reevaluate insect light reactions. In retrospect, the studies of Wellington et al. (1951) and Wellington (1955) are of considerable interest in that they involved sawfly and moth larvae. These investigations extended the phenomenon of polarized light sensitivity not only beyond the adult Hymenopterans, but also to organisms not possessing compound eyes. Both the sawfly and lepidopterous larvae were able to orient to the plane of polarized light, but the latter were more precise (probably because they have more stemmata of a somewhat different structure).

The question remains as to how insects can navigate unambiguously by polarized light when any given plane of polarization can be found at different points in the sky. Since insects can always view the zenith where

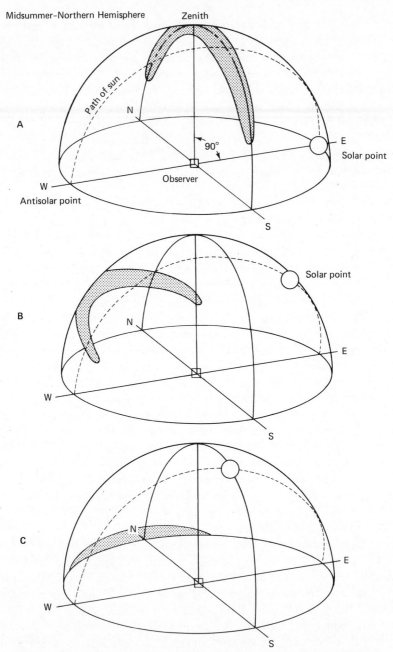

Figure 5–2. Changes in the location of the area of maximal polarization relative to the position of the sun and the observer. (A) When the sun is near the horizon, the area of maximum polarization is directly overhead. (B) About midmorning when the sun is about 45° above the horizon, the area of maximum polarization drops below the zenith to the north-west. (C) When the sun is overhead at noon, the area of maximum

the solar meridian extends at right angles to the plane of polarization, they can determine the path followed by the sun across the sky. However, additional cues would be necessary for an individual to distinguish between the two arcs of the meridian. Experimental evidence suggests that these cues are obtained from a view of a large area of sky. Wehner (1976) found that the honeybee could accurately communicate the direction to a food source by performing the waggle dance (Chapter 6) on a horizontal surface when a large portion of the sky was in view. However, when the dancers' view of the sky was limited to a small patch with a single plane of polarization, they danced alternately in two directions— one right and one wrong. It is interesting that the wrong direction was consistent but not that which would be predicted from the plane of polarization viewed.

The disc of the sun would seem to be a most suitable and reliable navigational reference point except that it may be obscured by clouds or may be out of sight during those periods when it is low in the sky. The ability of insects to view an area of polarized sky must provide a valuable alternative cue. Furthermore, Wehner (1978) found that the desert ant, *Cataglyphis,* oriented erratically when the sky was depolarized with a filter even though the sun was visible. He theorized that the position of the sun may not be identified as the brightest spot in the sky but rather as that area lacking polarized light; this could not be discerned when the entire sky was depolarized. Obviously, we have much more to learn about celestial navigation.

Over the past few years Wellington has been concentrating on the importance of polarized light in insect behavior rather than just how it is utilized. His findings have proved quite interesting. Wellington (1974a) noted that insects use two kinds of information for finding their way around familiar territory. They can traverse it directly by utilizing polarized light as a navigational aid or follow a zigzag pattern from one familiar landmark to another. Transients unfamiliar with the local terrain can be caught aloft by the passage of a cloud dense enough to obscure their view of the plane of polarized light and must settle until their solar reference is restored, whereas local residents can change their mode of orientation and remain active. Nevertheless, the pattern of activity and behavior was markedly affected by changes in the availability of polarized light to serve as a guide. For a period of time on either side of solar noon, polarized light cannot be seen in the overhead sky. During this period Wellington found that the passage of transients through the experimental

---

polarization is just above the northern horizon. (Reprinted, with permission from W. G. Wellington, *"A special light to steer by."* Drawing by H. Peter Loewer, *Natural History Magazine,* December 1974. © The American Museum of Natural History, 1974.)

area abated, as did the activity of residents in the open spaces; the resident flies, bees, wasps, and butterflies either restricted their activity to small localized territories or moved about from one route marker to the next.

Wellington (1974b) noted a parallel change in the navigation of foraging western bumblebees, *Bombus terricola*. When polarized light was present overhead, the workers would spiral upward from their foraging site and fly directly back to their nest in a straight line. When the polarized light overhead failed, the workers returned by a zigzag course dependent on landmarks. Coating portions of the compound eyes and the dorsal ocelli with an opaque substance revealed that the ocelli were important in navigation by polarized light during twilight when the light intensity was too low for the compound eyes to perceive landmarks.

Landmarks

It is obvious from the foregoing discussion that landmarks provide an important alternative to celestial navigational cues for a variety of insects. Beekeepers are familiar with the orientation flights of workers after a hive has been moved to a new locale; the workers fly around the hive in expanding circles, returning to the hive entrance and then flying off again as if to reconnoiter the immediate surroundings. When a number of hives are placed in a row in featureless terrain, the workers will tend to drift with the prevailing wind toward the hives at one end of the row. This can be prevented quite easily by painting a few hives different colors or adding banners or some other artificial landmarks. Many solitary wasps make similar circular orientation flights after having completed a subterranean nest.

Just what landmarks some of these insects use can be a mystery. Desert ants and the solitary wasps that nest in sand dunes occupy a seemingly featureless terrain, where shifting sand constantly changes even the subtle surface features. The Nobel prize winning behaviorist Tinbergen conducted an extensive study of homing in digger wasps, the essence of which is presented along with numerous other observations of wasp behavior in Howard Evans' *Wasp Farm*. Tinbergen noted that the bee-wolf, *Philanthus,* upon completion of its nest, flies upward and then circles the nest site several times. If identifying objects were removed from around the nest opening while the wasp was away foraging, the wasp would have difficulty finding the opening, and upon leaving again would engage in another study of the immediate vicinity—behavior not necessary when the identifying landmarks were left undisturbed. Tinbergen also found that moving landmarks would result in the returning wasps missing their nest openings.

I have watched *Ammophila* females make similar reconnoitering flights after completing a nest. However, these wasps disguise the nest opening before they leave and return with their prey on foot, often over a considerable distance. Evans (1973) described the return of a female that had just captured a caterpillar in his vegetable garden as follows: "The *Ammophila* proceeded straight down between two rows of peas with her caterpillar slung beneath her. When she reached the end of the garden, about twenty feet away, she made a right angle and followed a plow furrow for another five feet. Then she ascended the far side of the furrow and entered a patch of weeds where, with scarcely a hesitation, she dropped her caterpillar and began to dig."

The dune-inhabiting *Bembix* apparently are able to recognize landmarks consisting of no more than minor depressions in the sand. Some of these wasps hunt over considerable distances and return to their nests periodically even if their search for prey has been unsuccessful. It has been theorized that these wasps are able to memorize the location of larger and more distant objects and their relationship to the small ones near their nests.

Electromagnetic fields

Electromagnetic fields would seem to be an ideal navigational cue. An insect orienting to such a field could navigate equally well during both day and night, and travel long distances without need for the time compensation necessary in celestial navigation. There is evidence that insects can perceive electromagnetic fields. The honeybee, for example, responds to the earth's electromagnetic field in a way that incorporates systematic errors in the straight run of the waggle dance performed on a vertical comb. Lindauer and Martin (1968) were able to eliminate these errors by placing the dancing bees in an artificial magnetic field. Furthermore, a honeybee swarm transferred to a cylindrical container oriented their comb in the same direction as combs in the hive from which they were taken (Lindauer and Martin, 1968); this has been considered the result of magnetic field imprinting. Although Gould et al. (1978) discovered the presence of magnetic material in the abdominal region of honeybees, there is still no behavioral evidence that they use the perception of the magnetic field in their navigation.

Chemical trails

For social insects the establishment of chemical trails to aid in orientation to a food source and back to the nest provides a very effective aid to navigation. Such trails are used primarily by ants but also by a few species that forage on the wing. The termite, *Zootermopsis nevadensis*, also produces a trail pheromone (Hummel and Karlson, 1968). South

American bees of the genus *Trigona* are known to mark a trail from a food source by depositing a droplet of mandibular gland secretion every few meters on their way back to the hive (Lindauer, 1971).

Many ants establish one or more well-traveled trails between their nest and their foraging site. Sometimes these routes are so well trodden that they are easily followed once established. At least to begin with, the route back to the nest may be determined by celestial cues or by following landmarks, but often it consists of a well-marked pheromone trail. Chances are that any dense column of ants traveling back and forth across some featureless terrain such as an area of pavement is following a chemical trail. Often it is possible to demonstrate that such is the case by obliterating a section of the trail with soap and water. If a chemical trail exists, the ants arriving at either end of the obliterated portion will begin to mill about at random until several individuals reunite the divided column. The literature on the chemical trails of ants has grown rapidly in recent years and now includes species belonging to a variety of strategy groups. The trail pheromone is commonly a secretion of Dufour's gland, as it is in the harvester ant, *Pogonomyrmex badius* (Hölldobler, 1971a), and the fire ant, *Solenopsis saevissima* (Wilson, 1962).

In the fire ant, workers go out alone in search of food. Successful workers returning to the nest extrude their sting and leave streaks of pheromone along the ground. In this species the trail substance both activates and guides other foraging workers. However, the pheromone is highly volatile, and the trail declines to a nondetectable level in about 2 minutes. Since only ants returning to the nest with food add to the trail, when a food source is depleted recruitment ceases; thus foragers are not distracted by encounters with old useless trails (Wilson, 1962).

CONCLUSIONS

Studying the orientation and homing of insects within their home range or as they travel back and forth between their nests and places where they find their food is much easier than studying the navigation of long-range migrants over many miles of variable terrain. Mark and recapture experiments with migrant butterflies such as the monarch, *Danaus plexippus,* leave no doubt that these insects can and do fly up to 2000 miles. Not only do monarchs fly great distances, but after their southward migrations they end up at predictable locations that would seem to require rather precise navigation. Just how they accomplish this has not been demonstrated convincingly. The problem is that single individuals have not been followed over long distances, so the ability of migrants has been inferred from observations made as they pass through a series of localities along their route. Some butterflies and moths traverse specific mountain passes with seasonal regularity, whereas others seem to follow the coastline or other prominent topographic features. Certainly, insects are capable of using celestial cues, but, with the variable direction and velocity of winds they encounter along their way, how do they know when they have gone far enough?

Obviously, insects can find their way by perceiving and responding to a variety of cues including some, like polarized light, that have little meaning in terms of our own sensory systems. We should not assume that insects function in accord with human experiences, since their homing ability would certainly seem to surpass our own. Bees can communicate the distance between a nectar source and the hive by means of what Wilson (1971) called "a ritualized and miniaturized imitation of their journey." Ants can be trained to travel a specific distance from a source of food to their nest; when displaced from the nest area they will travel that distance in a straight line and then begin to search for the nest entrance. Many foraging species that hunt back and forth over a large area are capable of returning to their nests with considerable precision. Some behaviorists feel that insects employ kinesthetic memory (the ability to keep track of both the distance and directional changes on a journey and then to translate the information into a more direct return trip), but such feelings are supported mainly by negative evidence. Nevertheless, insects seem to be capable of navigational feats that we can perform only with the aid of a computer.

CHAPTER 6

Communication

With the exception of species like the desert locust and others that migrate en masse, the migratory behavior of insects typically leads to the separation, rather than aggregation, of individuals and introduces some associated problems. Mating often occurs before migration, in which case only the inseminated females migrate, but in many species various types of communication brings the sexes together after their migration is complete. In fact, communication plays an important role in the reproduction of almost all species. Communication also produces aggregations of individuals that enhance resource exploitation, overcome host resistance by mass colonization, increase survival, and in the case of social species stimulate cooperation between individuals.

Insects employ tactile, visual, auditory, and chemical methods of communication, and in many species a combination of methods is involved in patterns of behavior that fulfill a single biological function. For example, a special odor or sound may be produced to bring together a number of scattered individuals to mate and reproduce. Once in close proximity, pairing may result from visual recognition, and courtship behavior also may involve visual cues. Once paired, the male may induce the female to copulate by releasing a volatile aphrodisiac or by exciting her tactilely. Because so many of the behavioral sequences engaged in by insects involve more than one method of communication, as just illustrated, it is desirable to begin with a discussion of the basic means of communication before considering their integration and the functions they serve.

CHEMICAL COMMUNICATION

In the broadest sense, chemical communication among insects involves the transfer of information by way of the detection of chemicals present in the environment, particularly insect-produced compounds collectively called **pheromones**. Pheromones can be likened to hormones in that they are of a specific composition, produced by special glands, to be released at specific times. Furthermore, they inhibit or stimulate specific biological

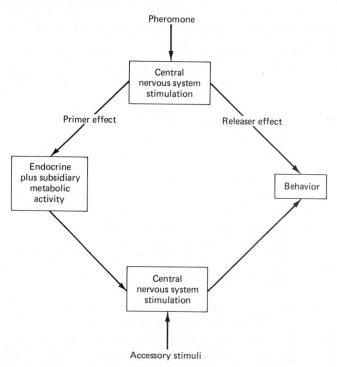

Figure 6–1. Schematic diagram of the direct and indirect influences that pheromones have on the behavior of insects. If a pheromone stimulates an immediate change in behavior, it is said to have a releaser effect. If it has a long-term physiological effect that later influences a response to some accessory stimulus, the pheromone is said to have a primer effect. (From "Pheromones" by E. O. Wilson, Copyright © 1963 by *Scientific American, Inc.* All rights reserved. In T. Eisner and E. O. Wilson, eds., *The Insects, Scientific American,* W. H. Freeman and Company, Publishers, San Francisco, 1977.)

functions. Whereas hormones coordinate the physiological and behavioral processes within the individual, pheromones coordinate the physiological and behavioral activities between individuals of the population. The rapidly expanding field of research involving the chemical control of insect behavior has been reviewed by a number of authors, including Karlson and Butenandt (1959), Butler (1967), Shorey (1973), and Shorey and McKelvey (1977). Pheromones are now thought to influence behavior, both directly and indirectly as illustrated in Figure 6–1, and most workers now use the terms **pheromone, allomone,** and **kairomone** defined in Table 6–1.

Insect pheromones can be variously grouped according to the kind of behavior or activity they coordinate (Table 6–2). Some pheromones function over long distances as sex attractants or as population aggregators, whereas others function only over short distances, as is true of some sex

Table 6–1. Description of Terms Currently Used in the Chemical Control of Insect Behavior

Pheromone	A chemical or mixture of chemicals that is released by one organism and induces a response by another individual of the same species; these chemicals can be either synthesized by the emitting organism or acquired intact from their food or other sources in their environment
Allomone	A chemical or mixture of chemicals that is released by one organism and induces a response, adaptively favorable to the emitter, by an individual of another species; most such chemicals would be defensive secretions
Kairomone	A chemical or mixture of chemicals that is released by one organism and induces a response, adaptively favorable to the recipient, by an individual of another species; many such chemicals act as host attractants

Source: H. H. Shorey and J. J. McKelvey, Jr., eds., *Chemical Control of Insect Behavior,* Interscience, John Wiley & Sons, Inc., New York, 1977.

stimulants. Consequently, pheromones can be important in stimulating reproductive behavior, producing aggregations at sources of food, stimulating mass defensive tactics, regulating population density, aiding navigation, and so on. Among social insects, pheromones are particularly important in coordinating members to perform all of the activities necessary for the survival of the colony as a whole.

Sex pheromones are employed by a wide variety of insects (see Jacobson, 1965) but probably have been most widely studied among the butterflies and the moths. In many species, the female produces a pheromone that attracts males to her location. Usually, the scent is released only during those periods of the day when the males are active and responsive but continues to be released throughout that period until mating has occurred. In species that mate several times, the female will release her pheromone for several days, thus ensuring the receipt of sperm from several males. In the moths, the pheromone molecules are perceived by thousands of olfactory receptors located on the large plumose antennae of the males. These special receptors (see Figure 1–5B) are capable of detecting a single pheromone molecule, and the excitation of only a few hundred receptors at one time by a pheromone concentration of only a few hundred molecules per cubic centimeter can cause the male moth to respond with a basic change in behavior, which may lead into a complete behavioral repertory (Schneider, 1974). When the antennal receptors of a resting male intercept the scent of a "calling" female, the male will often respond by vibrating its wings and then taking flight. The male will fly into the odor-bearing airstream until he locates the female or can no longer perceive her scent. If the male loses the scent, he will engage in a random flight that increases his chances of intercepting the trail again. In this way, widely separated males and females are effectively brought

Table 6–2. Categories of Behavior Modifying Chemicals Based on the Types of Behavior They Induce

Locomotory stimulant	A chemical that causes kineses that, in the absence of orientation cues, often cause animals to disperse from an area by *increasing* their speed of locomotion or appropriately affecting their rate of turning
Arrestant	A chemical that causes kineses reactions that, in the absence of orientation cues, often cause animals to aggregate near the chemical source by *decreasing* their speed of locomotion or appropriately affecting their rate of turning
Attractant	A chemical that causes an animal to make oriented movements toward its source
Repellent	A chemical that causes an animal to make oriented movements away from its source
Feeding, mating, or ovipositional stimulant	A chemical that elicits one of these behavioral reactions
Feeding, mating, or ovipositional deterrent	A chemical that inhibits one of these behavioral reactions

Source: V. G. Dethier, L. B. Brown, and C. N. Smith, The designation of chemicals in terms of the responses they elicit from insects, *J. Econ. Entomol.,* **53**:134–136, 1960.

together; in one experiment more than one quarter of the released males of the saturniid moth, *Arctias selene,* were able to locate caged females about 7 miles away.

Chemical communication between mates is particularly important for bisexual species in which the females are flightless. Such is the case with a number of moths belonging to the families Saturniidae and Lasiocampidae, in which the females are either extremely sluggish or have short, nonfunctional wings. In some of the tussock moths, for example, a newly emerged female sits on her empty cocoon, releases the pheromone that attracts a male, mates, and then lays her eggs in the immediate vicinity; the young larvae then migrate as described in Chapter 4.

Sex pheromones not only serve to bring males and females together but they may also cause some sexual excitation that facilitates copulation. Sometimes a low concentration of a pheromone will only agitate the male, but a higher concentration will stimulate oriented flight, and an even higher concentration will result in courtship and copulation (Traynier, 1968). In some species the females need a little stimulation as well, and the males produce a scent that acts as an aphrodisiac. For example, the male of the grayling butterfly, *Eumenis semele,* has a patch of special scent scales on the upper surface of the forewings, which are exposed to the female by the spreading of his wings during courtship. When the female encounters the pheromone, she is stimulated to engage in copulation.

Some insects release a mixture of chemicals that attract both sexes. This is common among the bark and ambrosia beetles, which use pheromones not only to attract members of the opposite sex but also to create an aggregation of individuals collectively large enough to overcome host resistance or, in the case of susceptible hosts, to maximize the utilization of contagiously distributed resources. In the genus *Dendroctonus,* the females initiate the attack of the host tree and subsequently release a sex-aggregating pheromone. In those genera such as *Ips,* in which the brood gallery is started by the male, the male produces the pheromone. As more beetles are attracted to the host and attack it, more pheromone is released, strengthening the stimulus. Obviously, such a procedure cannot go on indefinitely. As the host material approaches an optimal level of occupancy, the chemicals released by the host and the beetles change, and the attraction of new individuals slows down. Subsequently, the fully occupied host becomes unattractive as the production of the attractant compounds ceases, and the production of an antiaggregation pheromone, which acts as a deterrent to late arrivals, takes its place. Thus chemical communication can serve to optimize spacing and reduce intraspecific competition.

Many ant species lay chemical trails that assist them in traversing the distance between their nests and sources of food they have located. Ant trails are frequently marked by the release of small droplets of pheromone at regular intervals along the travelled route. In addition to marking the way, the released pheromones serve to recruit other individuals to gather food from the same source. As numerous ants travel back and forth over the same route, leaving spots of pheromone as they go, the spots blend together to form a continuous trail. The fact that ant trails involve chemical communication can be readily demonstrated by rather simple experiments involving the disruption of an established trail by replacing a portion of it with unmarked soil. It is also possible to create artificial trails with a crude extract of the ant's Dufour's gland, where the pheromone is produced. Termites mark their trails with a pheromone produced in special glands located in their abdomens. In the primitive species, an odor trail is used to recruit workers to damaged portions of nests that need to be repaired. In some more complex species, however, the same pheromone is used to recruit workers to food gathering, as in the ants. Some primitive bees also use spots of pheromone laid down at regular intervals to create a trail even though they forage on the wing (see Chapter 5).

Chemical communication serves a variety of other functions among nonsocial insects, both within and between species. Some parasites mark their prey with a pheromone that discourages oviposition by another female of the same species. For example, the braconid *Microphanurus* that parasitizes the eggs of the green vegetable bug, *Nezara,* marks each parasitized egg by depositing a pheromone with a circular motion of the tip of her abdomen. If another female locates the same bug egg mass, she will parasitize only those eggs that are not marked; in this way intraspecific competition is avoided. Some parasitic hymenopterans also use pheromones to mark the territory they have traversed in search of prey and thereby avoid repeating the search of the same area.

As one might expect, chemical communication is not restricted to members of the same species. There are many examples of insects using the chemical characteristics of other organisms as a means of identifying them as suitable hosts. Nowhere is this more apparent than in the relationship between insect herbivores and their food plants (to be discussed in Chapter 8). There are examples, however, that involve one species keying in on the pheromone communication of another. When working with bark beetle pheromones in the field, I have often seen large numbers of both parasites and predators attracted to natural and artificial sources of bark beetle attractants. Obviously, this kind of behavior is highly beneficial to species that would otherwise have to search at random for their prey.

Although in a relatively short time investigators have come to realize that chemical communication among insects is widespread and serves many functions, much remains to be learned. Workers investigating the chemical communication systems of pest species such as bark beetles are discovering that mixtures of different pheromones released in different concentrations elicit different responses. The integration of chemical communication with other methods of communication is even more complex, as we will discover a little later in this chapter.

AUDIO COMMUNICATION

Sound is also an effective means by which insects are known to communicate over a wide range of distances, but not over the long distances described for some pheromones. Sound is perceived by specialized mechanoreceptors, which may or may not be grouped together into ears. Since these receptors can detect any sound within their range of receptivity, sound itself does not necessarily provide a specific means of communicating any more than the detection of some common chemical does. However, some insects produce specific sound sequences or "songs" that form a functional parallel with the specific molecular mixtures of pheromones.

Insects may produce sound as a by-product of another activity, such as flight, by striking their substrate, by rubbing two parts of their body together, by activating a vibrating membrane, or by pulsing an airstream (Chapman, 1969).

The sound produced by the beating wings is used by some insects to locate other individuals and to bring the sexes together. A swarm of small flies is clearly audible to other individuals in the area. Males of the mosquito *Aedes aegypti* are attracted to the flight tone of sexually mature females, whereas immature females have a different tone to which the males do not respond (Jones, 1968).

A variety of insects produce sound by striking the substrate with some part of their body. Some grasshoppers strike the ground with their hind tibiae. Because both sexes engage in the activity, it probably serves to bring individuals together by way of a response to low frequency vibrations through the substrate. Individuals of the termite *Zootermopsis* rock back and forth on their middle legs so that the mandibles tap against the floor of their tunnel. This behavior is initiated by a disturbance near one part of the nest and warns other members of the colony to move elsewhere (Howse, 1964).

Stridulation is the term applied to the production of sound by the rubbing together of two body surfaces. The mechanism, which is used by many different insects, can be likened to running one's thumb nail down the teeth of a comb. Usually, a ridgelike structure, the **scraper**, on one part of the body is moved back and forth over a ridged surface, the **file**, on an adjacent part. Variations of this type of system are common among the grasshoppers, crickets, katydids, true bugs, and beetles. In the grasshoppers, a row of pegs along the inside of the hind femora is rubbed back and forth over the edge of the parchmentlike forewings. In katydids and crickets, the cubital vein of each forewing is toothed. When the wings are folded at rest, a ridge near the base of the left wing overlaps the file on the right wing. When these insects sing, their wings are repetitively opened and closed part way and the sound is produced by a vibration of the adjacent wing membrane (see Figure 6–2).

Beetles use many different parts of their well-sclerotized bodies as stridulatory organs, but the **elytra** are most commonly involved. Male bark beetles of the genus *Dendroctonus* have an abdominal scraper that is moved back and forth across an elytral file to produce a clearly audible chirping sound. The function of the sound is not understood but commonly occurs when the male is sitting in the entrance to a female's gallery while the host tree is under attack.

The production of sound by a muscle-driven membrane, or **tymbal**, occurs among a few moths and homopterans. The most thoroughly studied mechanism of this type is that of a well-known insect chorister, the cicada. In these insects, the first abdominal segment is highly modified for sound production. There are a pair of dorsolateral structures that resemble drumheads; each consists of a thin disc of cuticle, supported by a thicker cuticular rim. These discs form the tymbals, which are protected by tymbal covers composed of regular cuticle. Each tymbal gives rise to an apodeme on its inner surface, to which the tymbal muscles are attached. When the muscles contract and cause an inward buckling of the tymbals, there is an associated click. When the muscles relax, the tymbals return to their normal shape and produce a second click. The succession of clicks associated with a cycle of rapid muscle contraction and relaxation produces the familiar trill of the cicada.

The only known case of sound produced by a pulsed airstream occurs in the moth genus *Acherontia,* which draws air in through its proboscis by dilation of the pharynx. The in-rushing air causes a flaplike epipharynx to vibrate and to produce a pulsed airstream and a related low-pitched sound. This is followed by a high-pitched whistle as the air is expelled.

Insect-produced sounds can serve a variety of functions, such as the stridulatory sounds that accompany defensive displays or occur in response to agitation. I have a cerambycid beetle in my garden that produces a loud startling buzz when touched. More important from a behavioral standpoint, however, are sounds used for communication within species.

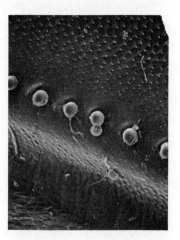

Figure 6–2. (A) A scanning electron micrograph of the inner surface of the hind femur of a grasshopper showing the structure of the stridulatory file. (B) Diagram of the stridulatory apparatus of the cricket *Acheta.* (Redrawn from R. E. Snodgrass, 1935, *Principles of Insect Morphology,* McGraw-Hill Book Company, New York.)

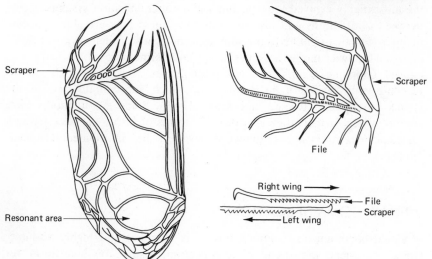

Functionally, acoustic behavior is most important to activities associated with territoriality and reproduction and, consequently, plays a role in speciation. Some specific examples will therefore be presented when these topics are dealt with later on; however, Haskell (1964) presented a general outline of acoustic signals in reproductive behavior, which are paraphrased here: A calling song produced by either the male or the female stimulates locomotor activity and brings the sexes together, provided the recipient of the song is ready to court. A courtship song which may be divided into several distinct accoustic signals, such as a serenade and nuptual song, may lead to the actual joining of the genitalia. During copulation the male may continue to calm the female with a copulatory song. Following copulation the pair may go their separate ways or there may be some postcopulatory singing, which in crickets apparently discourages dislodgement of the spermatophore.

A spectacular form of audio communication among insects is the seemingly synchronized singing called chorusing. Such behavior occurs in populations of male cicadas, crickets, and katydids, and results in the attraction of females and sometimes recruits additional males to the chorus. A number of hypotheses suggested to explain this behavior failed to recognize the current view that natural selection operates principally at the level of the individual. Alexander (1975) discussed the problems of interpreting chorusing behavior and concluded that the phonoresponses involved can be explained primarily as competitive interactions between neighboring males.

Audio communication also seems to be important in predator-prey systems involving insects. According to Haskell (1964), insect predators can locate prey both on land and in the water by responding to the vibrations created by their prey; some hymenopterous parasites are believed to locate hosts lying beneath the substrate in a similar manner. One of the most highly evolved interspecific audio communication systems occurs between some moths and insectivorous bats; these bats are able to locate flying insects by orienting to the source of the echos of their own ultrasonic chirps. The owlet moths have evolved a countermeasure in the form of a pair of tympanic organs that detect the hunting bat's cries. The moths respond with evasive flight maneuvers, which substantially reduce the chance of capture (Roeder and Treat, 1961). Some arctiid moths have gone one step further; they also produce trains of ultrasonic clicks, which for some reason cause bats to swerve away at the last moment (Roeder, 1965).

VISUAL COMMUNICATION

Visual communication among insects is common and highly variable. Unlike chemical and sound communication, visual communication requires a direct line of sight between individuals and is only effective over relatively short distances. Within its effective range, vision can be important as a means of bringing individuals together and also in the sometimes elaborate patterns of courtship behavior.

The females of mayflies, caddisflies, and midges are attracted visually to swarms of males engaged in dancelike up-an-down flights in compact groups. Male butterflies frequently pursue appropriately colored females and can be persuaded to chase pieces of colored paper moved in a manner that imitates a female's flight. Once butterflies have paired, visual displays become an important part of the courtship behavior without which copulation may not occur.

The most spectacular visual displays encountered among the insects involve luminescence (see Lloyd, 1971). In some insects, luminescence is incidental and serves no specific communicative function. Some collembolans, for example, give off light as a by-product of their metabolism, as do many bacteria; insect larvae infected by such bacteria may also appear luminescent. A survey of luminous insects, as well as a review of the physiology and biochemistry of bioluminescence, is presented by McElroy (1964).

The most brightly luminous insects are beetles belonging to the families Elateridae, Phengodidae, and Lampyridae; variously known as fireflies, lightning bugs, and glowworms. The lampyrid, *Photinus pyralis,* which is common in the southern United States, has been studied extensively. Both adults and larvae have light-producing organs—the larvae and females have a single pair, whereas the males have two pairs. The flashing signals produced by these organs seem to be a species specific means of communication widely recognized as a mechanism that brings the sexes together.

The behavior of *Photinus pyralis* has been described by McDermott and Mast (in Buck, 1948) as follows: "At dusk the male and female emerge from the grass. The male flies about 50 centimeters above the ground and emits a single, short flash at regular intervals. The female climbs some slight eminence such as a blade of grass and perches there. She ordinarily does not fly at all and she never flashes spontaneously as does the male but only in response to a flash of light which is produced by the male. If a male flashes within a radius of 3 or 4 meters of the female she usually responds after a short interval by flashing. The male then turns directly toward her in his course and soon glows again. Following this the female again responds by glowing and the male again apparently takes his bearings, turns and directs his course towards her. This exchange of signals is repeated usually not more than 5 or 10 times until the male reaches the female and mates with her."

In *P. pyralis* the female flashes about 2 seconds after perceiving the flash of the male, and this apparently aids in sexual recognition. A number of individuals flashing within a small area at the same time would seem to produce a confusing situation, but perhaps it is unscrambled, at least in part, by the critical response distance displayed by the males relative to a flashing female. Synchronized flashing by male *P. pyralis* is rare, but it is reportedly commonplace among some tropical species. How this synchronous behavior is triggered is unclear, and it raises the same fundamental questions regarding selection as does the synchronous chorusing mentioned earlier.

TACTILE COMMUNICATION

Communication by touch obviously can occur only after other means of communication bring individuals together. Nevertheless, insects engage in rather elaborate forms of contact behavior for courtship and sexual stimulation. In *Drosophila,* for example, the male is attracted to the female visually, but final species recognition results when the male taps the female with his forelegs. In many species the female is not immediately receptive to the male's copulatory advances and will only allow the male to mount after some appropriate foreplay. In a few insects the male appeases the female with an offer of food and then copulates with her while she eats his offering. Male scorpion flies of the genus *Panorpa* (Mecoptera) secrete droplets of saliva that harden and serve as a snack for the female during copulation. The males of other species appease their mates with seeds or nonedible objects such as a brightly colored petal.

The appeasement behavior of males would seem to guard against exciting a female to respond to a mate as a potential attacker or prey. In the mantids the male carefully approaches the female from behind, over a protracted period, but even then the female responds by grabbing him violently with her forelegs and then proceeds to remove his head with her mandibles during copulation. The outcome, however, is vigorous copulatory activity resulting from the severing of the male's subesophageal ganglion.

THE INTEGRATION OF COMMUNICATION METHODS

In the foregoing discussion the four main methods of communication used by insects have been treated separately. However, we must realize that they are not used independently. Each has its special attributes. Odors, for instance, are carried great distances on the wind and, if detectable in low concentrations, can provide a way by which insects beyond the range of sight or hearing can communicate effectively. At closer range, other forms of relaying information may be more appropriate. Furthermore, biological functions are frequently fulfilled by way of complex behavior patterns composed of a sequence of responses. Such is often the case with those functions served by communication.

For mating to occur, individuals may have to come together over considerable distances. The danger of attack by natural enemies is considerably lessened when individuals can communicate while they remain concealed. Odor and sound would seem to serve this strategy well. Once individuals have been able to gain proximity, visual cues would become useful as a means of species recognition. Finally, tactile stimulation can serve to provide the excitation required before the female submits to copulation.

A behavioral sequence of the type just described, involving several types of communication, is relatively simple compared to the communication that must take place among the subsocial and social insects. In truly social species, the survival of entire colonies depends upon the integration of the relatively simple individual patterns of communication and response into a coordinated system of mass behavioral phenomena. An examination of communication within a colony of honeybees therefore serves as probably the best example of cooperative behavior mediated by various forms of communication.

The coordination of life and behavior within a beehive involves communication of each of the types discussed. Since social bees return to their nest at regular intervals, mechanisms for long-distance communication have been replaced largely, but not totally, by elaborate communication in an around the nest site. Although visual, sound, and tactile communication occurs between members of a hive, chemical communication predominates. In addition to chemical communication, as discussed previously, bees exploit chemical signals in the form of special substances in the food that is exchanged between individuals. (The glands of a honeybee that produce behavior-mediating chemicals are shown in Figure 6–3.) The following discussion, therefore, will be organized according to the functions served by communication rather than by the method employed. The discussion will of necessity be shorter than the subject justifies, but interested students are referred to the outstanding treatment by E. O. Wilson (1971) for further details.

Wilson divides social communication functionally into alarm and assembly, recruitment, recognition, food exchange, grooming, and group effects. A brief examination of the kinds of communication used in the regulation and coordination of these functions in only a single species should provide some perspective of the complex nature of this subject over the entire range of social insects.

Alarm and assembly

Alarm and assembly can be used for both defensive and foraging purposes. An initial sting by a honeybee may provoke other bees in the vicinity to become aggressive. Because the worker's sting is barbed, it catches in the victim's skin, and, as the bee attempts to fly away, the

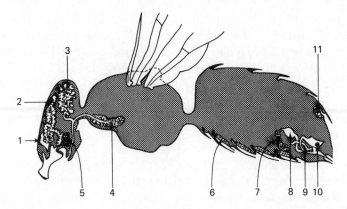

Figure 6–3. Chemical producing glands of the worker honeybee *Apis mellifera:* (1) mandibular glands, (2) hypopharyngeal gland, (3) head labial gland, (4) thoracic labial gland, (5) hypostomal gland, (6) wax glands, (7) poison gland, (8) vesicle of the poison gland, (9) Dufour's gland, (10) Koschevnikov's gland, (11) Nasanov's gland. (Redrawn from E. O. Wilson, 1971, *The Insect Societies,* The Belknap Press of Harvard University Press, Cambridge, Mass., with the permission of the author and Harvard University Press.)

poison gland and **Dufour's gland** are often left behind. The act of stinging releases the compound isoamyl acetate, produced by secretory cells that line the sting pouch (Shearer and Boch, 1965). This highly volatile substance attracts other bees to the source, and some subsequent stinging may occur. However, it does not stimulate the release of the substance by other bees not engaged in stinging, so, as the danger passes and the alarm pheromone dissipates, behavior returns to normal.

Much more important in honeybee behavior is the chemical communication that leads to other forms of assembly. Worker bees release a mixture of chemicals produced by their **Nasanov glands**. The substance is often released by bees located near the hive entrance during swarming, and when new food sources are first located. Individuals that have been isolated from their nest mates for a period of time will also release the scent as a means of reestablishing contact.

The scent continuously laid down around the entrance to the hive assists the foraging bees to locate the nest on their return. However, this hive odor does not serve to distinguish one colony from another, as was once thought. During swarming, the Nasanov substance stimulates the assembly of workers that ultimately leads to the familiar cluster of bees around their queen.

Another assembly pheromone consists of a group of chemicals, commonly called **queen substance**, produced by the mandibular glands of the queen. This secretion and another from **Koschevnikov's gland** are at least partially responsible for the formation of the cluster of "court bees" that constantly surrounds the queen. Also, when a colony swarms, the workers are attracted to the queen in flight and follow her trail of evaporating pheromone to the settling site located by the scout bees. Once the queen settles, she releases another mandibular gland secretion that tends to settle the nearby workers. The first group of settled workers dispense their own Nasanov secretion, which stimulates the rest of the swarm to cluster.

Recruitment

Recruitment involves the gathering together of nest mates at a particular place for purposes of applying a joint effort to a specific task, such as nest construction or food retrieval. The most elementary form of recruitment communication in the honeybee involves the recognition of food sources from the scent that adheres to the bodies of foraging bees and the nectar they regurgitate upon return to the hive. When a particular food source is abundant and near the hive, this simple form of communication is quite adequate. In fact, investigators have increased pollination by training the bees on sugar syrup tainted with the odor of the crop (von Frisch, 1967).

Ants forage for food on the ground and so are able to leave well-marked trails to recruit their nest mates to the task of exploiting a food source. Bees, on the other hand, forage on the wing, so they must have some other means of communicating the exact location of food to their colleagues. Some bees do lay odor trails, however. As discussed previously, the South American genus *Triogona,* having established a course from the hive to a food source, will stop every few meters on the homeward flight and deposit a droplet of mandibular gland secretion at each point. Other bees then follow the odor trail (Lindauer, 1971). The honeybee employs a rudimentary form of odor trail used for short-distance orientation in the immediate vicinity of the hive. The trail is laid down by workers returning to the hive over a short distance on foot; the chemical is referred to as the **footprint pheromone**.

The honeybee communicates the distance and the direction to good foraging sites by an elaborate method of communication widely known as the **bee dance** (von Frisch, 1967). Different races behave slightly differently, but usually, if the source of pollen or nectar is fairly close to the hive, returning workers perform the **round dance**, as shown in Figure 6–4. As indicated in the figure, a worker that has just returned from a successful foraging trip penetrates the hive to where there are other field bees and engages in an excited circular pattern of running. Other field bees follow and in the course of the dance pick up odor information about the source of food. This simple form of the bee dance recruits other workers to search for pollen or nectar of a certain kind in the immediate vicinity of the hive.

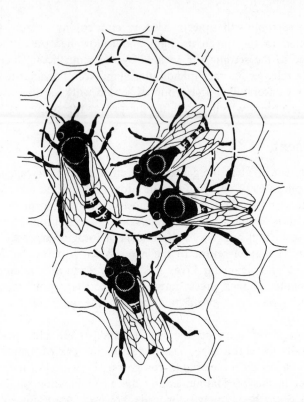

Figure 6–4. The round dance of the scout honeybee that is thought to communicate the presence of a nectar source near the hive. (Reprinted from Karl von Frisch: *Bees: Their Vision, Chemical Senses, and Language.* Copyright © 1950, © 1971 by Cornell University. Used by permission of the publisher, Cornell University Press.)

If the food source is more than 50 meters or so from the hive, the information transmitted by the round dance would be of little use. As the distance to the foraging site increases, the round dance becomes modified by the incorporation into it of a straight run, during which the performing worker waggles its abdomen from side to side. The new pattern of behavior that results is called the **waggle dance** (Figure 6–5). After much painstaking study, von Frisch was able to decode this remarkable means of communication.

When a scout bee returns to the hive after having made several successful trips to and from a forage site a considerable distance away, she penetrates deep into the hive, regurgitates nectar from her crop, and begins to perform the waggle dance on one of the vertical honeycombs. The sun is used as the key reference point for the communication of direction. If the forage site is located on a line between the hive and the sun, the straight run of the dance will be up the comb. If the direction is directly away from the sun, the straight run will be down the comb.

Figure 6–5. The waggle dance of the scout honeybee that is believed to communicate both the direction and the distance to nectar sources away from the hive (see text for details). (Reprinted from Karl von Frisch: *Bees: Their Vision, Chemical Senses, and Language.* Copyright © 1950, © 1971 by Cornell University. Used by permission of the publisher, Cornell University Press.)

The straight run is followed by a circle to the right, another straight run followed by a circle to the left, and so on. Likewise, if the forage site is located 30° to the right of the sun, the straight run of the dance will be performed at an angle 30° to the right of the vertical (Figure 6–6). The compound eyes of the honeybee are so sensitive to ultraviolet radiation that they can detect the position of the sun and communicate the direction to forage sites, even on lightly overcast days, and navigate using polarized light.

The communication of the direction to more-distant food sources is highly beneficial, but foraging efficiency would clearly be enhanced by some communication of distance as well. This information is provided by the duration of the straight run and the speed with which the dance is performed. During the straight run, the dancing bee waggles her abdomen from side to side with a frequency of about 13 to 15 vibrations per second, at the same time producing an audible buzzing sound by vibrating her wings. By observing the dance of scout bees returning from feeding platforms placed at different distances from the hive, von Frisch was able to decode the dance language. The further the food from the hive, the

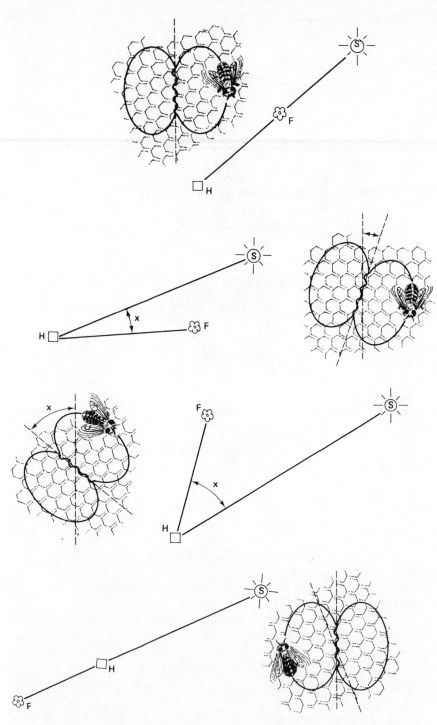

Figure 6–6. Diagrams showing the orientation of the straight run of the waggle dance relative to the position of the hive (H), the forage site (F), and the sun(S).

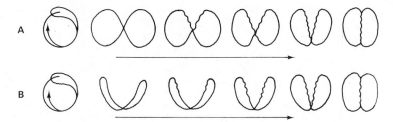

Figure 6–7. Changes in the round dance that occur as the distance between the hive and the foraging site increases. (A) Changes characteristic of the Austrian strain. (B) Changes characteristic of the Italian strain. (Adapted from K. von Frisch, 1967, *The Dance Language and Orientation of Bees,* trans. by L. E. Chadwick, The Belknap Press of Harvard University Press, Cambridge, Mass.)

longer was the duration of the wagging portion of the dance and the fewer were the number of complete cycles of the dance per unit of time.

Apparently, the duration of the straight run is not based on the absolute distance to the foraging site but on the energy that must be expended to get there. If the trip involves a flight up a steep slope or against the wind, the straight run performed in the hive will be longer. Furthermore, different genetic races of honeybee have incorporated variations into their dance that have been referred to as dialects (von Frisch, 1962).

For example, the transition from the round dance to the waggle dance differs between the Austrian variety (Figure 6–7A) and the Italian variety (Figure 6–7B); the flattened figure-eight pattern or **sickle dance** of the Italian race is used when the nectar source is at an intermediate distance from the hive, the opening of the sickle facing the source of food. A study of the dance behavior of three wild species of *Apis,* in addition to the Austrian and Italian races of the honeybee (von Frisch, 1962), revealed substantial differences in the correlation between the rapidity of the wagging motion and the distance to the nectar source (Figure 6–8). In *Apis mellifera* there is even a difference between the dialects of the varieties, which results in misinterpretation of the dance language in a mixed group. Experiments with colonies containing both Austrian and Italian workers revealed that the Austrian strain consistently overestimated the distance communicated by Italian dancers, whereas Italian foragers underestimated the Austrian dancers (von Frisch, 1967).

The waggle dance is a truly fascinating example of insect communication, but it is even more amazing than meets the eye. As Wilson (1971) so clearly stated, it is " . . . a signal constructed from a ritualized and minaturized imitation of the journey that the signalling bee has taken in the past and upon which some of its sister bees are about to embark." The workers in the hive are able to rehearse their flight in miniature before they set out, much as we do when we trace a proposed plane trip on a map.

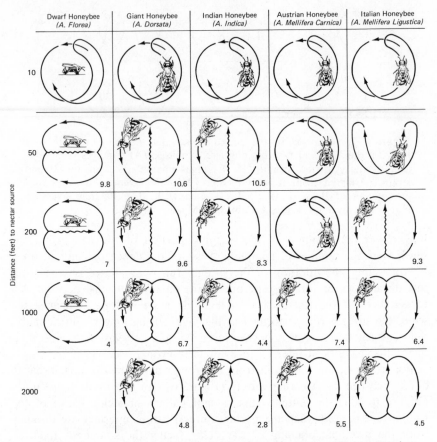

Figure 6–8. Dialects in the dance language of several different species and strains of the honeybee genus *Apis.* The dwarf honeybee dances on a horizontal surface whereas the others all dance on a vertical surface. The more rapidly the wagging portion of the dance is performed, the shorter the distance to the forage site. The values in the corner of each square represent the number of wagging runs in 15 seconds for each distance for each variety of bee. (From "Dialects in the language of the bees" by Karl von Frisch. Copyright © 1962 by *Scientific American, Inc.* All rights reserved. In T. Eisner and E. O. Wilson, eds., *The Insects, Scientific American,* W. H. Freeman and Company, Publishers, San Francisco, 1977.)

In 1967, Wenner and Johnson challenged von Frisch's interpretation of the bee dance as the major means of communication for foraging. They contended that the bees could adequately achieve recruitment with the odor of the nectar brought to the hive and the recruitment pheromone left by foragers in the field (Wenner et al., 1969). This stimulated additional research. The results indicate that the two hypotheses, although in conflict, are not mutually exclusive, as pointed out by Gould (1975). Gould concluded that von Frisch and Wenner were simply examining different parts of the same process. The elaborate dance language is of great value in the rapid recruitment of foragers to remote isolated patches of food, but, when an extensive supply of food is available, recruits can use odor alone. As Gould pointed out, the honeybee evolved in areas where a means of exploiting spatially and temporally discontinuous food sources that were also sought by other species would have been a significant advantage.

Honeybees engage in other dance variations that have yet to be studied as thoroughly as the waggle dance. One called the buzzing run initiates swarming. During the warm part of a suitable day, one or several workers begin running through the colony in an excited zigzag pattern, vibrating their wings and abdomens as they go. The sound produced is quite distinctive and may be part of the total pattern of communication. The dance is highly contagious, and soon many workers join in. Within minutes, the bees near the hive entrance rush out and are followed by a mass of workers and the absconding queen. Wilson believes that this form of communication may be unique in that the initial signal stimulates others to produce the same signal in the form of a chain reaction.

Recognition

Recognition is an extremely important part of the communication system of social groups. In a bee colony, for example, there is so much activity involving the comings and goings of the field workers that an inability to distinguish friend from foe could be a major problem. The recognition of nest mates seems to be very casual, but intruders of other species are quickly and violently expelled. The reaction of honeybees to individuals of the same species, but from a different hive, seems to vary considerably in relation to the possible significance of the intrusion. During periods when the nectar flow is strong, the accidental arrival of a field bee from another colony will probably be accepted. But when nectar is scarce, strangers are repelled and robber bees meet with violent rejection. Each colony does not have an inherently distinctive odor as once thought, but each colony does become recognizable by the odor of the pollen and nectar the bees are collecting. During a major nectar flow, when all the hives are working a common food source, there is probably little difference between them. On the other hand, during hard times, each hive may have a very distinctive odor determined by the nectar source each has located.

The recognition of castes is also important among social insects. The queen honeybee is clearly distinguished from all other members of the colony and is treated with apparent respect. The workers not only recognize the queen on the basis of the odor substances she secretes but are also able to determine when she is becoming reproductively ineffective.

After the old queen leaves a colony with a swarm, the first virgin queen to emerge communicates audibly with other young queens who have not yet emerged. If the first queen to emerge killed all the others, the colony could be left queenless by an accident on her nuptial flight. By producing a chirping sound called "piping," the first queen delays the emergence of her sisters, who chirp back from within their cells. This has been interpreted as a means of delaying the mortal combat that must occur whenever there are two queens in a colony until after the first queen has had a chance to mate.

Group effects

Group effects in the honeybee are obvious from the incredible control that a viable queen exerts over the functioning of her colony. As long as a healthy queen is present to give off the appropriate signals, the workers contribute unerringly to her needs, those of her offspring, and the colony as a whole. If a queen is removed, the organized behavior of the workers gives way to disorganized restlessness in a matter of minutes. Within a few hours, workers will draw several worker cells into enlarged cells suitable for rearing a replacement queen. After a few days, some of the workers experience ovarian development and actually lay infertile eggs, which develop into drones. The presence of queen substance apparently acts to inhibit queen cell construction and reproduction, probably by suppressing the activity of certain endocrine glands of the workers. This raises the question of how the workers are stimulated to produce young queens to facilitate the reproduction of the colony by swarming. Butler (1960) found the amount of queen substance produced by queens during the normal swarming season to be about one quarter of that of queens of nonswarming colonies. Thus the queen clearly exerts her control over all individuals that comprise a colony; yet, she in turn is influenced in her behavior and egg laying by the workers that tend to her needs.

INTERSPECIFIC COMMUNICATION

A variety of communication types may also be involved in the behavioral interaction between species. This is well illustrated by the relationship between some ant species and other insects that they treat like guests in their nests. The guest species, called **myrmecophiles**, live in the ants' nests, where they may be groomed and cared for by their hosts and even permitted to eat the ant larvae. Ants, like the honeybee previously discussed, have an elaborate communication system that coordinates nest construction, food gathering, brood rearing, and defense of the colony. The fact that the ants allow some alien species full access to the benefits of their society suggests that the guests have, in the words of Hölldobler (1971b) " . . . broken the ants' code, that is, attained the ability to 'speak' the ants' language which involves a diversity of visual, mechanical and chemical cues." Hölldobler reports that the larvae of the rove beetle,

Atemeles pubicollis, which live in nests of *Formica polyctena,* produce a
secretion that imitates the pheromone that ant larvae emit to stimulate
the brood-keeping behavior of the adults. The beetle larvae also imitate
the begging behavior of ant larvae, which mechanically stimulates a brood-
keeping adult ant to regurgitate a droplet of food (Figure 6–9).

Figure 6–9. Chemical communication between a beetle larva and its
ant host. (A) The position of the glands of the beetle larva that produce
an attractant substance. (B) The worker ant is attracted to the beetle
larva and (C) is stimulated to engage in grooming. (D) The tactile stimulation causes the beetle larva to rear up and, if the larva makes mouth-to-
mouth contact with the ant, (E) the ant regurgitates a droplet of liquid
food. (From "Communication between ants and their guests" by B. Hölldobler. Copyright © 1971 by *Scientific American, Inc.* All rights reserved. In T. Eisner and E. O. Wilson, eds., *The Insects, Scientific American,* W. H. Freeman and Company, Publishers, San Francisco, 1977.)

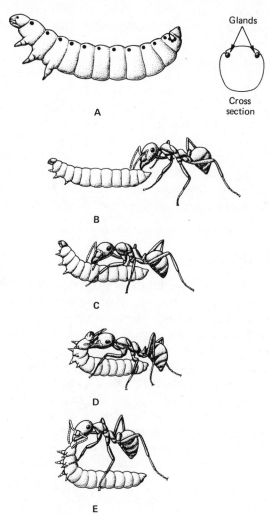

CHAPTER 7

Reproduction

Reproductive behavior in most bisexual insect species can be subdivided into mate location (the coming together of individuals of the opposite sex), courtship and copulation, and oviposition. In many species the courtship phase may also involve some aggressive behavior among rivals and the establishment and defense of territories. Oviposition may be followed by some level of brood care. In this chapter I will discuss only those aspects of reproductive behavior that precede oviposition, since the laying of eggs often involves host selection behavior, which is discussed in Chapter 8. Brood care will be considered in Chapters 10 and 13.

As indicated in the previous chapter, communication in nonsocial species serves reproduction more than any other function. Pheromones are of great importance in the formation of aggregations of both sexes in the drawing of one sex to another; sex pheromones are particularly widespread among nocturnal species. Both sound and visual cues are effective over shorter distances, but, with the exception of biolumi-nescence, visual cues are used principally by diurnal species.

MATE LOCATION

The simplest form of mate location involves either active hunting or waiting for a mate to pass nearby. In some hemipterans, for example, the male approaches any moving object of appropriate size and determines whether or not it is conspecific only at close range (Markl and Lindauer, 1965). In many species the male will not respond to an object unless the visual stimulus is more specific. For example, in the Odonata, dragon-fly males of the family Lestidae will only approach objects that fly like a dragonfly and have transparent wings, whereas males of the damselfly, genus *Calopteryx,* respond only to objects of appropriate size, flight rhythm, wing color, and wing transparency (Markl and Lindauer, 1965). The specific visual cues and mate location behavior of the bioluminescent fireflies, *Photinus pyralis,* were presented in the previous chapter.

The mayflies, caddisflies, stoneflies, many of the more primitive Diptera (Nematocera), members of a few families of higher flies, and a few hymenopterans engage in group reproductive behavior called swarming (not to be confused with the process of colony division in honeybees). Swarming, often involving only males, usually results from the common response of a number of individuals to a visual marker, rather than an aggregation response to other individuals. The markers can be highly variable, ranging from a small patch of sunlight amongst the shade of vegetation or the silhouette of a branch against the sky, to a prominent geographic feature such as a hilltop. The flight of swarming insects is balanced against the wind so that the group remains more or less in one place. Within the swarm, the individuals often move up and down and from side to side, giving the appearance of a dance motion. The swarming behavior is claimed to increase sexual excitement and probably strengthens the signals received by females in the vicinity; females are often seen to fly into or near a swarm, where they may be captured by a male and mated. Swarming in Diptera is reviewed by Downes (1969).

The common response of both the males and females of some species to a visual marker constitutes an efficient method of bringing together the sexes of a dispersed population (Downes, 1958). The silhouette of a hill often forms an optical marker toward which insects orient from a distance, finally ending up at the peak. Some butterflies, ants, and strong-flying flies (particularly tabanids and syrphids) congregate on summits (Chapman, 1954). As is often the case, males outnumber females in the vicinity of swarms, probably because the males tend to remain whereas the females once mated, leave. Although olfactory trail markers led female bees to the swarming place of the males, in some species such as the common honeybee, drones swarm in the same place year after year without any prior experience, and virgin queens will also seek these swarming places, apparently by using visual cues (Ruttner and Ruttner, 1965).

Once some form of behavior has brought the sexes together, the recognition of a conspecific of the opposite sex must precede courtship and copulation although courtship itself may consist partially of recognition behavior. In some species the male will simply attempt copulation with any object that attracts it. In many, however, conspecifics recognize each other by various means, and the sex is recognized by some specific response after meeting. At close range, visual, olfactory, and tactile cues all can be important. For instance, color patterns seem to be important to recognition among butterflies and some flies; female mosquitoes that enter a swarm of males can be identified by the sound of their wing beat (Jones, 1968); and drones recognize virgin queens by the odor of their mandibular gland secretion (Lindauer, 1967).

RIVALRY AND TERRITORIALITY

When males outnumber females in an area, the competition for mates can result in rival behavior between males of the same species in addition to aggressive responses to males of other species. Probably, males actively searching for mates or engaged in courtship enter into clashes far more often than actually observed because such behavior could be exhibited anywhere within the species' range. Nonterritorial rival fighting occurs among males of the silverfish, and mantid males will respond to other males in a manner similar to their attack on a prey. Various grasshopper males respond to another calling male with a rival song, which apparently reduces the time males might otherwise engage in courtship singing. Most observations of rivalry fights have been made in what can be considered a territory, even though insects have not generally been considered to be strongly territorial organisms. Markl and Lindauer (1965) define a territory as the area in which a male, ready to mate, awaits a female, or an area to be used for oviposition. In addition, some social species stake out and protect foraging territories.

The establishment and defense of territories have been studied most extensively in the dragonflies and crickets. Dragonflies and damselflies belonging to several genera establish aerial territories that fan out over a body of water from a daytime perch—often the stem of a plant that overhangs the water's edge. The established male responds to an intruder by rushing toward it. In genera such as *Calopteryx* that are sexually dimorphic, the territorial male will attack other males and attempt to copulate with females. In genera such as *Libellula* that lack sexual dimorphism, the territorial male will attempt copulation with both males and females (Corbet, Longfield, and Moore, 1960). Johnson (1964) indicated that male intruders may just be chased away, but mating attempts may lead to an in-flight scuffle with the victor taking charge of the territory. A territory may change hands several times in the course of a day or be maintained for several days by the same male. A diagram of typical adult patterns of behavior is presented in Figure 7–1.

I recently observed a similar behavior pattern in the syrphid fly *Volucella* sp. Lone males use a corner of a deck on my house as a visual marker. The fly in possession of the territory hovers, maintaing an almost stationary position typical of many swarming dipterans. Periodically, a fly hovering on station will drop its legs and a few seconds later dart off quickly in the direction of another fly. Usually, the two flies engage in aerial maneuvers reminiscent of a World War II dog fight; thereafter one of the flies returns to the original station and begins hovering. It has not been possible to determine whether the original territorial male or the interloper is the usual winner, but Collett and Land (1975) studied similar behavior in the syrphid fly, *Eristalis,* and found that following a chase the

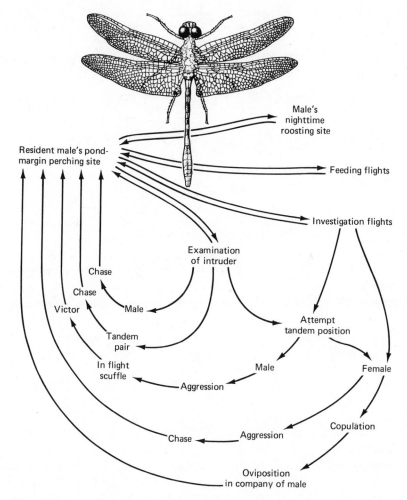

Figure 7–1. Diagrammatic schema of territorial and reproductive behavior of the dragonfly *Libellula saturata* (Uhler). (Redrawn from D. L. Johnson, 1964, Master's thesis, San Diego State University, San Diego, Calif.)

same fly returned to the hovering station. Occasionally, when *Volucella* flew out after an intruder, the two flies would become coupled together and fall to the ground. I presumed these intruders to be females that the territorial males captured and attempted to copulate with, but I have been unable to collect a pair for examination.

The aggressive and territorial behavior of male field crickets has been studied in detail by Alexander (1961). When two males encounter each other, a variety of interactions can ensue, but the end result is almost always the retreat of one of the individuals (Figure 7–2). Alexander recognized five levels of encounter based upon the intensity or kind of

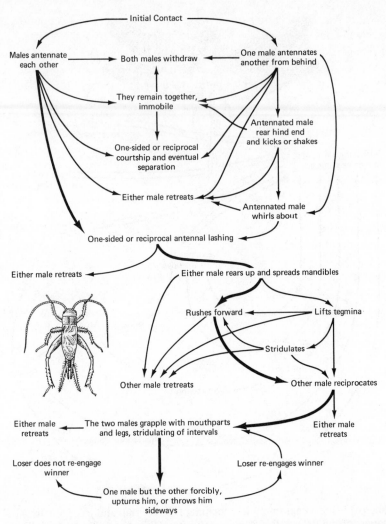

Figure 7-2. Diagrammatic schema of the typical interactions that occur between adult male field crickets. The dark arrows indicate the usual sequence of responses when two aggressive males meet. (Adapted from R. D. Alexander, 1961, *Behavior*, **17**: 130-223.)

aggression exhibited. The most intense encounters occurred between evenly matched males; crickets apparently establish a hierarchy on the basis of age, recency of copulation, territoriality, and recent success in fighting. Fighting usually begins with antennal lashing, but if the combat is not broken off, the fight will proceed to sparring with the forelegs, kicking, and even wrestling with locked mandibles; however, in spite of the apparent violence of the battle, mutilation rarely occurs.

When male crickets occupy burrows, as some species do quite commonly, encounters between males are less frequent. The occupation of a burrow enables males to establish a territory and resist the aggression of individuals that would be dominant in encounters in the open. Alexander (1961) suggests that the reduction of male-male encounters afforded by burrow ocupation frees the occupant to spend more time singing his mating song that attracts potential mates. The occupancy of burrows also disperses the male population over a larger area and thereby enables females to orient more easily toward a single singing male.

Davies (1978) described the ritualized combat over mating territories displayed by the males of the speckled wood butterfly, *Pararge aegeria,* which defend sunlit spots on the forest floor. Whenever an interloper intrudes upon an occupied territory, it is challenged by the occupant. The two butterflies then make a short spiral flight upward into the forest canopy, after which one returns to the sunlit spot (Figure 7–3). Davies conducted experiments that demonstrated it is invariably the owner, rather than the individual of superior size and strength, that returns to the spot. Apparently, the spiral flight communicates that the spot is occupied, and that fact is seemingly accepted readily by the interloper. Such behavior would assure the full occupation of courtship territories with minimal expenditure of energy and without the infliction of injury, but it raises a question as to how territories change hands or how vigorous interlopers obtain mating spots in a fully occupied area of the forest.

COURTSHIP

Once conspecifics of differing sex have been drawn close to each other, there would be a selective advantage to behavior patterns that would suppress aggressive or escape responses and lead to sexual receptiveness. Courtship serves both of these purposes and includes an array of behaviors that ranges from a slightly modified attack response to ritualized attack, appeasement, sexual stimulation, and finally, copulation.

Courtship involves all forms of communication. Pheromones are used in numerous species for both recognition and sexual excitation. Males of some beetles will attempt to mate with a piece of paper treated with female sex pheromone, while ignoring an actual female held captive under glass nearby. During the courtship of the grayling butterfly, *Eumenis semele,* the male spreads its wings and bows toward the female so that her antennae come in contact with scent patches on his forewings (Figure 7–4). Many grasshoppers move about the female in an excited manner while singing a courtship song prior to attempting copulation. Visual displays are also common and sometimes spectacular, as are the aerial displays of some butterflies (Figure 7–5) and the bioluminescent flashing of fireflies. Tactile stimulation, though often not obvious, is also important

Resident chases away interloper

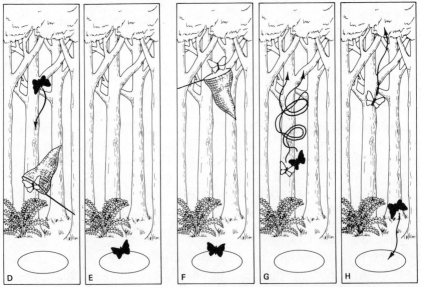

Interloper occupies vacant spot   New resident chases away previous resident

Figure 7–3. Diagram of the ritualized combat over mating territories displayed by males of the speckled wood butterfly, *Pararge aegeria*. (A–C) The usual pattern of behavior displayed when an interloper (black) approaches a resident (white). (D–G) The experimental design used to demonstrate that the resident always retains control of the territory. (Redrawn from N. B. Davies, 1978, with the author's permission, *Anim. Behav.* **26:** 138–147.)

Position of
scent patch

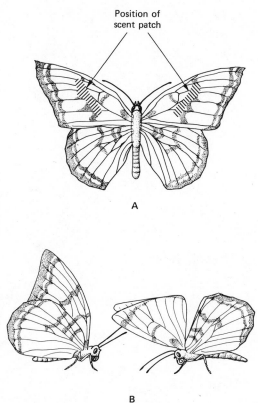

A

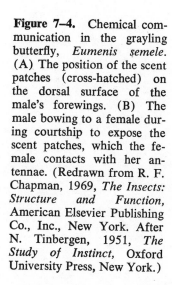

B

Figure 7–4. Chemical communication in the grayling butterfly, *Eumenis semele*. (A) The position of the scent patches (cross-hatched) on the dorsal surface of the male's forewings. (B) The male bowing to a female during courtship to expose the scent patches, which the female contacts with her antennae. (Redrawn from R. F. Chapman, 1969, *The Insects: Structure and Function*, American Elsevier Publishing Co., Inc., New York. After N. Tinbergen, 1951, *The Study of Instinct*, Oxford University Press, New York.)

and widespread. In some species the male offers the female an object or some food, which apparently increases the female's receptiveness. The male of the seed-feeding bug, *Stilbcoris*, usually approaches the female with a seed impaled on his beak. If the female accepts the seed, she will insert her beak into it, at which time the male moves closer, grasps her, and begins copulation while she continues to feed; males without seeds are rejected. A variety of insects, including cockroaches, scorpionflies, wasps, and flies, regurgitate material that the female consumes prior to or during mating (Chapman, 1969).

The courting of some species involves a long sequence of behavioral acts by the male to which the female may or may not respond. Sometimes the male's fixed action pattern consists of a long chain of actions that must be performed in its entirety once begun, but in other species the male progresses through the entire sequence only if he receives appropriate acceptance responses from the female at key points along the way. Such sequences have been studied in a number of insects, but two examples from opposite ends of the evolutionary spectrum of the insects will serve as examples.

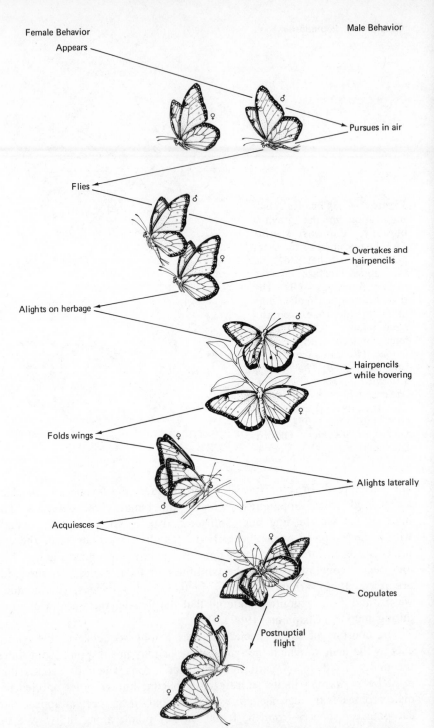

Female Behavior

Appears

Flies

Alights on herbage

Folds wings

Acquiesces

Male Behavior

Pursues in air

Overtakes and hairpencils

Hairpencils while hovering

Alights laterally

Copulates

Postnuptial flight

Figure 7–5. Courtship behavior of queen butterflies, showing the response chain resulting from an encounter between a male and a female. (After L. P. Brower, J. V. Z. Brower, and E. P. Cranston, 1965, *Zoologica,* **50:** 1–39, with permission.)

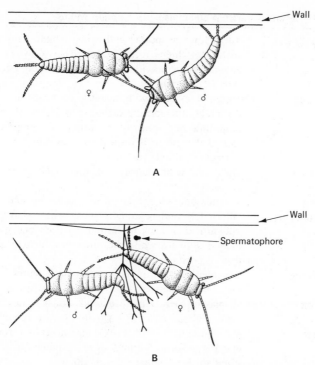

Figure 7–6. Courtship behavior of the silverfish, *Lepisma saccharina.* (A) The male approaches the female from the front and then (B) induces her to pass under a main silk thread (heavy line) attached to the wall and floor by establishing a number of secondary guiding threads (fine lines). The system of threads leads the female to the spermatophore deposited earlier by the male. (Redrawn from H. Sturm, 1956, *Tierpsychol.,* **13:** 1–2.)

The silverfish, *Lepisma saccharina,* engages in complicated precopulatory behavior in which the male produces a package of sperm (spermatophore) and then leads the female to it with a series of silk threads that restrict her movement. This so-called leading and bridling behavior (Figure 7–6) culminates with the male pushing the female's ovipositor into contact with the spermatophore (Sturm, 1956).

The courtship behavior of flies belonging to the genus *Drosophila* involves the most elaborate set of signals of any of the mating sequences studied thus far. A simplified catalogue of signals commonly used by male *Drosophila* is presented in Table 7–1 and the female responses are presented in Table 7–2.

Drosophila adults are attracted by odor to decomposing and fermenting plant material, particularly fruits, where they feed on the bacteria and yeasts and in many cases mate and oviposit. The commonality of this food odor response helps to bring conspecifics of both sexes together, but also results in the formation of mixed species aggregations. Although

Table 7–1. Major Courtship Elements of *Drosophila* Males

Orientation	♂ turns toward another individual, slightly raising body
Tapping	♂ lifts, straightens foreleg, and strikes downward against other individual; almost invariably occurs at start of courtship
Wing vibration	Oriented ♂ extends wing nearset ♀'s head, then vibrates it rapidly; interspecific variation in degree of extension, amplitude and speed of vibration, and angle of extended vane with respect to substrate, i.e., horizontal, tilted, vertical
Wing flicking	♂ flicks one wing sharply out, then back to resting position
Wing waving	♂ extends one wing, then slowly waves up and down
Wing semaphoring	♂ alternately and repeatedly flicks wings sharply outward, then back to resting position; one wing is moved outward while the other is returning to the resting position
Wing scissoring	♂ repeatedly and rapidly extends both wings horizontally outward and back to the resting position
Leg vibration	♂ displaying at rear of ♀ extends and vibrates forelegs against abdomen of ♀, usually against venter
Leg rubbing	♂ displaying at rear of ♀ extends and rubs forelegs to and fro against abdomen of ♀, usually venter, but some species against sides or dorsum
Circling	♂ periodically circles about ♀, facing her as he moves, often from rear to front and back, sometimes completely about female
Licking	♂ opens labellar lobes, extends proboscis, and licks ♀ genitalia; ♂ may either lick intermittently and repeatedly or continuously for a prolonged period; some Hawaiian species have modified labellar lobes which grasp ♀'s genitalia
Mounting	♂ curls tip of abdomen under and forward, rears upward, thrusts head under ♀'s wings or between her spread wings, grasps her body with his fore and mid legs, and attempts intromission
Countersignaling	Leg striking, kicking, wing-vibration signals resulting from one male attempting to court another male

Source: H. T. Spieth, Courtship behavior in *Drosophila, Ann. Rev. Entomol.*, **19:** 385–405, 1974.

there is some separation of species resulting from species-specific responses to time of day and environmental conditions, several species are likely to be present at the same place at the same time.

According to Spieth (1974) the females must feed steadily to fulfill the nutritional demands of egg production, whereas the males feed for only short periods, after which they turn their attention to mating. Since

Table 7–2. The Major Elements of the Courtship Behavior of *Drosophila* Females

Female acceptance behavior

Wing spreading	Female spreads both wings outward and upward and holds them extended until male mounts
Genital spreading	Female slightly droops tip of abdomen, slightly extrudes genitalia, and spreads ovipositors apart
Ovipositor extrusion	Female extrudes ovpositor posteriorly; restricted to Hawaiian species

Female repelling behavior

Decamping	Female breaks contact with courting male by running, jumping, or flying
Kicking	Female kicks vigorously backward with hind legs, striking face when he courts at her rear
Fluttering wings	Female rapidly flutters wings in movements of small amplitude; often occurs when male initiates tapping action
Abdomen elevation	Female elevates abdomen, thus raising tip high above substrate and inhibits male courtship such as kicking and leg vibration; often observed when male attempts to court feeding female; may be accompanied by extrusion
Abdomen depression	Female depresses tip of abdomen and wing tips against substrate; inhibits male courtship actions such as kicking and leg vibration
Extrusion	Female extends and elongates tip of abdomen, thus exposing articulating membranes around genital sclerites and simultaneously directs tip of her abdomen toward male face, usually causing male to turn quickly away and engage in cleaning behavior

Source: H. T. Spieth, Courtship behavior in *Drosophila, Ann. Rev. Entomol.,* **19**:385–405, 1974.

the males cannot visually distinguish between similar appearing species or conspecific males from females, they approach any passing individual with roughly the right conformation. Consequently, a system of signals has evolved that Spieth concludes has the advantage of providing the males with a means of ascertaining the receptiveness of females with a minimum expenditure of time and energy, the unreceptive females with a way to repel males without interrupting their feeding, and receptive females with an opportunity to sample several males thereby bringing sexual selection into operation.

In *Drosophila subobscura,* for example, the courtship ritual consists of a series of sexually alternating stimulation and response steps (Figure

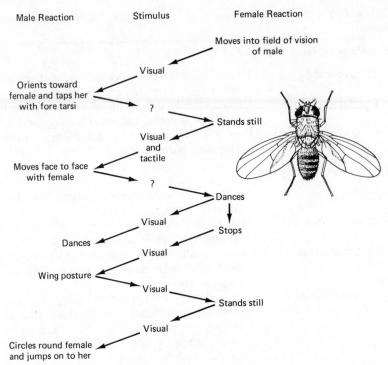

Male Reaction   Stimulus   Female Reaction

Figure 7–7. The sequence of stimulation and response steps that characterize the courtship behavior of *Drosophila subobscura.* (Adapted from R. G. Brown, 1966, *Behaviour,* **25**: 281–323.)

7–7). The male begins by facing the female and tapping her with his forelegs. If the male perceives the appropriate visual and tactile stimuli from the female, he will extend his proboscis and move to a face-to-face position. The male then taps the female's head and both begin a dance that consists of a series of side steps while still aligned face to face. As the dance progresses, the male raises his wings at right angles to the body with the leading edge tilted downward. The female responds by ceasing to dance. The male then circles his mate and finally jumps upon her and attempts copulation (Brown, 1966).

In most insects, successful courtship usually culminates in copulation. Various species-specific positions are employed (Figure 7–8), and the duration of coupling may range from seconds to hours. Postcopulatory behavior is also highly variable. In many species the pair separates immediately, and the female begins to search for an oviposition site whereas the male goes off in search of another mate. In others the pair remains together for some time. In some species the male continues to provide a food offering, which prevents the female from eating the spermatophore he has deposited, whereas in others, like the mantids, the female may

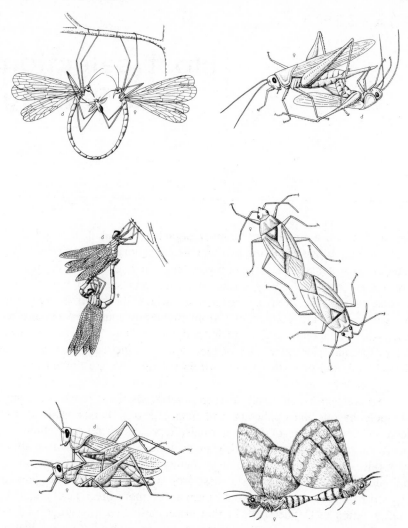

Figure 7–8. Various positions assumed by insects during copulation. (Redrawn from various sources.)

actually devour her mate. The females of a few species such as the screwworm, *Cochliomyia hominivorax,* mate only once, but most mate several times, although they are usually unresponsive to the advances of males for some time after each copulation. In a variety of species belonging to different orders, the male may remain with the female during oviposition. Some dragonflies, for example, fly with their mate as she deposits her eggs while flying over the surface of the breeding pond. In a sizable array of species the male actually cooperates in the rearing of the young, as we will see in Chapter 10.

CHAPTER 8

Host selection
and feeding

Insects feed on almost every form of organic material, living and dead, coal and natural gas being the two notable exceptions. Some species, often referred to as generalists, feed on a very wide range of foods—the roaches are a good example. Many species of scavengers are also rather omnivorous, in that the decaying material on which they feed is comprised of both animal and plant matter from numerous sources. Most insects, however, are more specific in their food relationships and restrict themselves to some particular kind of food (plant tissue, sap, nectar, animal flesh, blood, and so forth.) Often, within the food type, a limited number of species serve as hosts.

Several terms commonly used to describe the food range of animals are polyphagous, oligophagous, and monophagous. **Polyphagous** species feed on a wide range of foods normally representing several families of plants or animals, whereas **oligophagous** species feed on a narrower range of foods, often from within a single family. However, the separation of polyphagy from oligophagy is at best arbitrary. **Monophagy**—feeding restricted to a single species—would seem to be more precise, although the interpretation of Dethier (1947) that insects attracted to a group of food species by a common chemical stimulus are monophagous seems logical. Monophagy in the narrow sense is fairly limited, but the opposite extreme, **pantophagy**, probably never occurs.

When we view the over-all pattern of food preferences among the insects, there is a clear indication that some mechanisms exist that have divided the species among the food resources available. In the case of all insects that feed on other living species, one important mechanism is coevolution. Few multicellular, terrestrial, or freshwater species of plants and animals are immune to insect attack; the chinaberry tree, *Melia azedarack*, may be an exception. Conversely, there is probably no plant species attacked by all of the herbivorous insects that share its geographical distribution.

Insects and their hosts have been engaged in the process of food allocation by natural selection for a very long time. The host species disadvantaged by insect attack have had to evolve protective or defensive mechanisms in order to survive the attack and/or competition from other species. When a host species evolves some form of insect-feeding deterrent, the associated insect species, in essence, faces a reduction in its food supply. If it is to survive, the insect must also adjust and often does so by evolving a mechanism that, at least temporarily, overcomes the host's defense. This ongoing scheme of attack and counterattack has played a major role in the evolution of insects and their hosts. One important outcome of this process has been the reduction in the number of general food relationships and an increase in some degree of food specificity or preference.

The features of both the host species and its insect associate that bring about some change in their relationship must be inheritable so that these traits can be passed on from generation to generation and the degree of benefit they convey can be tested by natural selection. Often the changes are enzymatic or physiological, but behavioral adjustments also occur that reduce the chance of the insect's losing contact with its host during necessary periods of displacement.

Coevolution and other ecological mechanisms such as competition, which affect the way various resources are apportioned among the species of an ecosystem, are generally the same for all types of food relationships. However, the details of the stimulus-response patterns involved in the selection of plants by herbivores, insect hosts and prey by parasites and predators, and vertebrate hosts by blood feeders are somewhat different.

The host selection behavior of insects is complicated further by the fact that many species display some degree of adult-larval divergence in their food habits. In those species, such as the grasshoppers and many true bugs, the adults and their offspring feed side by side and display the same stimulus-response patterns in their plant host preference. In many species, however, the adults feed on an entirely different kind of food than do the young. Such is the case for most biting flies, a variety of aquatic species, most lepidopterans, many hymenopterans (particularly the entomophagous parasites), and some beetles. For these species, either the adults and young must display different patterns of food-finding and recognition behavior or the female must select the larval food as part of her oviposition behavior. In the latter group, food location, at least for the larvae, is never a problem, since, if the female does her job, they are surrounded by food from the time they hatch. If the female errs in the deposition of her eggs, her larvae die of starvation and, in the process, weed out any aberrant genes that would be detrimental to the species' established food relationship (Thornsteinson, 1960). Whether or not the surrounding food is satisfactory may still be ascertained by the larvae, however. In the case of the less discriminatory feeders, such as filter-feeding aquatic lavae, there is little or no problem. For some plant feeders, however, the larvae may have to select young foliage over old or locate a suitable feeding site.

Studies of the behavior of insect groups with different trophic relationships have had a somewhat different focus, and our understanding of the host selection process has not advanced to the same point in each group. Therefore, I will consider phytophagous insects, entomophagous insects, and blood-feeders separately.

HOST SELECTION BY PHYTOPHAGOUS INSECTS

Roughly one half of the species of insects feeds on plants. A relative few, like the electric buck moth, *Hemileuca electra,* which feeds exclusively on flat top buckwheat, *Eriogonum fasciculatum,* feed on a single plant species; probably none feed on all of the plants that occur within their range. Most feed upon the members of a single family or a few related families, as, for example, the white cabbage butterflies (Pieridae) that eat members of the mustard family. Some, like the gypsy moth, *Porthetria dispar,* which feeds on more than 200 species of plants from a wide range of families, have more catholic tastes, but may display some definite preferences (Campbell, 1979).

Insects use visual, chemical (olfactory and gustatory), and tactile cues in the location and recognition of the plants on which they feed. Whereas olfaction is an effective means of long distance communication among insects, most available evidence indicates that it functions most commonly over only short distances in host selection (Schoonhoven, 1968; Chapman and Kennedy, 1976). Unfortunately, most of the studies of olfactory preferences have been conducted under the unnatural conditions of still air in laboratory olfactometers of variable design. There have been only a few reports of insects in the field flying considerable distances from downwind to host odor sources (e.g., McMullen and Atkins, 1962), but of course, the fact that visual cues and other orientation mechanisms might have been operating cannot be ruled out.

The response of phytophagous insects to the color of vegetation seems to play an important part in food location but not in the discrimination of host species. As Thornsteinson (1960) suggested, the narrow spectrum of the light reflected from green plants would probably prohibit the decisive recognition of food plants by color. A number of phytophagous insects including aphids, leafhoppers, and certain beetles are strongly attracted to yellow and yellow-green, and it seems likely that these pests can locate crops by the color contrast between them and the surrounding native vegetation (see Kieckhefer et al., 1976; Roach and Agee, 1972). The color of flowers also attracts bees from a distance, although selection may be made at close range on the basis of fragrance; finally, visual cues provided by nectar guides may lead the bee to the location of the nectar (Manning, 1956).

Visual form can also aid in plant host location and selection. Bumblebees orient to the form of inconspicuous flowers (Manning, 1956). Some bark beetles orient toward vertical silhouettes in preference to horizontal ones, and the white pine weevil, *Pissodes strobi,* shows a preference for the longer, larger-diameter terminal leaders in a pine forest (VanderSar and Borden, 1977).

Tactile stimuli, in addition to contact chemoreception, also play a role in host selection, but probably to a lesser extent than visual and chemical cues. The surface characteristics of bark and foliage influence the choice of both feeding and oviposition sites. Bark and ambrosia beetles often display a preference for rough bark over smooth bark areas when initiating their brood galleries. The cereal leaf beetle prefers smooth-leafed wheats over varieties that are pubescent (Haynes, 1973), and some leaf-edge-feeding caterpillars that do not normally feed on holly leaves will do so when the sharp points are removed (Ehrlich and Raven, 1967).

The role of chemical stimuli in the selection of plant hosts by insects has garnered most of the research attention, perhaps because the manipulation of odor cues would have greater practical application. The premature leap from limited knowledge to sweeping generalizations about the mechanisms of insect plant host selection created considerable controversy, which in turn stimulated further investigation.

A few of the earlier ideas are now of little more than historical interest. These ideas have been refuted as often as they have been supported by experiments that frequently leave something to be desired in a design sense. The generalization that healthy plants are immune plants is a gross oversimplification. Although it is often true that healthy plants are able to sustain insect attack better than weakened plants, this does not constitute immunity in terms of a change in insect behavior. Furthermore, there are many insects that show a preference for young foliage over old, as well as irrigated and fertilized plants over what might be considered less healthy plants. Likewise, Hopkins' host selection principle, which states that the conditioning of larvae to a particular food plant determines the oviposition preference displayed in the adult stage, has been both confirmed and rejected experimentally numerous times.

The examination of some of the other principles of plant host selection has resulted in more significant contributions concerning both the nature of the chemical stimuli involved and the behavioral mechanisms involved. Obviously, these aspects are tightly interrelated.

One group of workers influenced by Fraenkel proposed that insects choose their host plants exclusively by responding selectively to plant secondary compounds, such as alkaloids, glycosides, and aromatic oils, that have no nutritional value (see Fraenkel, 1953). Fraenkel (1959) summarized his theory as follows: "The food specificity of insects is based solely on the presence or absence of these odd compounds in plants, which serve as repellents to insects in general and as attractants to those few which feed on each plant species."

Fraenkel's theory would seem to fit well with the concept of the co-evolution of insect herbivores and plants. It has been suggested that the feeding of herbivores was the selective pressure that led certain plants, particularly the angiosperms, to evolve chemical defenses in the form of metabolic by-products (odd substances). The herbivores would, in turn, have experienced a decline in their food supply, and some species are thought to have responded to this selective pressure by evolving some countermeasure; any species with a means of overcoming the plant's defensive mechanism would at least temporarily obtain an advantage over competing herbivores.

One might then envision that changes in the behavior of the insect might not only have helped to overcome a plant's defensive strategy but also to exploit it by using the odd substances as a means of host recognition (Ehrlich and Raven, 1964). In other words, an allomone would become a kairomone. Such a pattern of coevolution seems highly plausible, but Dethier (1970) cautions that the coevolution of plants and their herbivores is not a limited system. He suggests that the strength of selective pressures, such as competitors and the physical environment, might be changed by the feeding of herbivores. There is ample evidence that some insects exert a powerful selective pressure on their plant host, but most have a lesser impact. It would seem, therefore, that we should adopt the broader view of Dethier. If plants evolved odd substances as part of their defensive weaponry, several behavioral responses by insects are possible. The substances that may act as repellents, inhibitors (Chapman, 1974), or oviposition or feeding deterrents (Lundgren, 1975) to a range of herbivores might become chemical cues that attract and stimulate feeding by a few herbivores or have no effect at all. That is, if a herbivore was unable to detect a plant's odd substance(s), it would neither be repelled nor deterred (Jermy, 1966).

Other workers have proposed that insects select their food plants on the basis of nutrient value and respond to feeding stimulants such as sugar, lipids, and amino acids. Since these compounds are widespread but not recognized readily except by contact chemoreception, questions have been raised concerning their role in host selection. Clearly, some compromise theory is needed; the dual discrimination theory of Kennedy and Booth (1951), which suggests that selection is made on the basis of both odd substances and nutrients, seems more realistic. However, the wide variety of insect-host relationships (Jermy, 1976) seems to preclude any generality.

Some insects, like the aphids studied for many years by Kennedy, not only seem able to discriminate between host plants and even leaves within a species but are also able to discriminate between alternate host species at different times of the year. Others, like the gypsy moth mentioned earlier, are able to feed on a wide range of host plants. However, this does not mean they are indiscriminate, and their more catholic taste would seem well adapted to their habit of dispersing as first instar larvae, or vice versa.

On the other hand, there are insects that seem bound to a limited number of plant hosts by the latter's odd substances. Several insect species, including the cabbage butterflies, are attracted to the Cruciferae by the presence of mustard oil glycosides, but even here sapid nutrients are equally essential to food selection (Thornsteinson, 1960). Other species cited as examples of plant host specificity do not respond positively to any identifiable secondary compound. For example, the Colorado potato beetle, *Leptinotarsa decemlineata,* which feeds on the potato family (Solanaceae), apparently does not recognize or respond to the alkaloids that act as deterrents to other species; primary substances such as sugars and amino acids along with the secondary compounds seem to act synergistically in the initiation and sustaining of feeding by the beetle.

Dethier (1970) proposed that the diversity of feeding habits displayed by phytophagous insects might be expected on the basis of the fact that "Behavioral changes can be both rapid and reversible (unlike physiological and morphological changes—M.D.A.) and thus constitute a formidably effective response to whatever evolutionary innovations plants essay." He also proposes that the coevolution of insects and plants need not have resulted from a sequence of attacks and counterattacks, but rather from random mutations in both the neural systems of insects and the chemical systems of plants. He suggests that this could account for the fact that some plants evolved deterrents to insect feeding before the phytophagous insects evolved, and might be the reason why some plants that could well serve as food for many insect species do not.

HOST SELECTION BY BLOOD-FEEDING INSECTS

The number of insects that feed on blood are few in comparison to those that feed on plants. Although there are a few blood feeders scattered throughout a number of orders (i.e., Thysanoptera, Mecoptera, Lepidoptera, and Hymenoptera), the majority are true bugs (Hemiptera), sucking lice (Anoplura), biting flies (Diptera), and fleas (Siphonaptera). Members of these latter groups extract blood from a variety of invertebrates—particularly other insects—and vertebrates belonging to all classes. The flightless sucking lice are mainly associated with birds, and spend their entire life cycle on their host or within the host's nest. The fleas, also flightless, are often closely associated with their host in a physical sense and tend to be fairly species-specific. Among the hemipterans and biting flies, there exists a wide range in the degree of host specificity, as well as the dependence on blood as food. Most of the research attention, however, has been focused upon those species that have man as a principal host, namely, a few true bugs (especially *Rhodnius* and *Triatoma*), mosquitoes, black flies, and tsetse flies.

In spite of the comparative lack of species diversity among blood feeding insects and the concerted effort to gain an understanding of the factors involved in the selection of individual human hosts, generalities seem to be as elusive here as in the area of plant host selection. Useful recent reviews include Hocking (1971) and Friend and Smith (1977).

As in phytophagous insects, blood-feeding species seem to employ the full spectrum of their sensory capabilities in host location. If a generality can be made regarding differences in stimuli used by plant and blood feeders, it is that vision—particularly color vision—is less important, and thermoreception is clearly more important, for the latter group. In the area of chemoreception, it is once again difficult to determine the relative importance of nonnutritive odd substances, phagostimulants, and nutrients.

Visual stimuli are most important among daytime active blood feeders, as one might expect. Tabanid flies seem more responsive to movement than to other visual characteristics, whereas blackflies, although responsive to movement, are attracted to dark silhouettes with a matte rather than glossy finish; the shape of solid geometric silhouettes did not seem to effect attraction although the flies did land most frequently near the corners or points of the targets (Bradbury and Bennett, 1974).

Most blood-sucking flies do not display a high level of host specificity and frequently feed on nectar and water in addition to blood; the tendency of some insects to extract blood from commonly available hosts can be misleading. Humans are a suitable and sometimes favored host of blood feeders, perhaps because of the readily available, naked feeding surface. The fact that insect bites cause considerable discomfort and often form the basis of disease transmission has stimulated a search for the factor or factors that make one host more or less desirable than another.

One often hears reports suggesting that some people are bitten by insects more frequently than others, and scientists have latched on to this fact and used it as the starting point of host selection research. Concerning host selection by mosquitoes, Hocking (1971) states: "Every imaginable part of the human body and its products must by now have been fractionated in every possible way in search for a specific chemical attractant." Although many compounds including amino acids, sex hormones, and some metabolic by-products have been found to be attractive to mosquitoes, none are as attractive as an intact individual. Brown (1966) found moisture to be the most powerful single attractant to mosquitoes, with warmth next in importance. Carbon dioxide seemed to be an important activator. Several estrogens and over half of 26 amino acids tested were also attractive. However, there is some uncertainty as to the role played by the latter group of substances, as they are not particularly volatile and do not occur naturally in a pure state.

The importance of thermal stimuli to host location by mosquitoes has been demonstrated by a number of investigators, but Brown (1966) and Hocking (1971) conclude that the evidence supports the detection of convection currents (which can be either warm or cool relative to the ambient air) over radiant heat. The attractiveness of convectional currents may be enhanced by the odors they carry (Khan et al., 1968), and vice versa. However, in the case of *Rhodnius, Triatoma,* and the bed bug *Cimex,* the thermal response is probably to a temperature gradient.

In general, the location and selection of hosts by blood feeding insects seems no less complex than that of phytophagous species. There is a multiplicity of possible stimuli, many of which, like temperature and volatile chemicals, are fundamentally inseparable. For this reason laboratory experiments that examine responses to a single stimulus may produce spurious results.

HOST SELECTION BY ENTOMOPHAGOUS INSECTS

The food habits of entomophagous predators and parasites (parasitoids) are highly variable. Generally, predators are less discriminatory than parasites, but there are some predators, such as the coccinellid beetle, *Vedalia cardinalis,* which preys almost exclusively on the cottony cushion scale, *Icerya purchasi,* which are prey-specific. Predators also tend to consume prey in both adult and larval stages, although various combinations occur. For example, syrphid fly adults are primarily nectar feeders, whereas the larvae of some species are active predators; some flies of other families are scavengers as larvae and predators as adults. Parasites, on the other hand, are more likely to display greater host specificity and adult-larval feeding divergence is common. Most of the parasitic hymenoptera, for example, are parasitic as larvae and nectar-pollen feeders as adults.

Predators tend to capture their prey by either sitting and waiting for them to come within reach or by active pursuit—a few, like the larvae of antlions, construct traps (Figure 8–1). In many species vision plays a predominant role in prey capture. Ambushers, like preying mantids, respond to movement and will strike at any object of appropriate size when the stimulation of specific groups of omatidia (the fovea) indicate that it is within range (Maldonado and Barros-Pita, 1970). Pursuit-type preda-

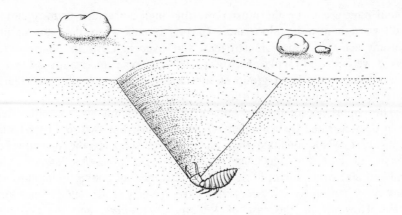

Figure 8–1. Section through the pit trap constructed by the larva of an ant lion showing the position of the larva lying in wait for its prey.

tors will respond to most moving objects of appropriate size, although the more prey-specific species may subsequently select prey on the basis of tactile or chemical stimulation. The wasp *Philanthus* will pursue a variety of prey but will only capture bees or prey tainted with the odor of bees (Chapman, 1969). Mechanical stimuli may also be important. Water striders, for example, locate prey trapped in the surface film by orienting to the ripples that radiate outward as the prey struggles to free itself. Aphid-eating coccinellids will search more or less at random by visually scanning their vicinity; once an aphid has been captured and consumed, the beetle will engage in a localized search involving frequent turning movements. This type of behavior is particularly effective for predators that feed on gregarious and more or less sedentary prey.

Most entomophagous parasites are carnivorous only as larvae. Although a few species in several different orders have a specialized, actively hunting, first instar larva that locates a host for itself, most depend on the capacity of the female to oviposit on or in a suitable host. How the female locates and selects her oviposition site is as variable and complex as host recognition by plant and blood feeding species.

In some species, the female does not search for the host but simply oviposits in the host's habitat. For instance, some tachinid fly parasites lay their eggs on the foliage of plants that serve as the food for their caterpillar hosts; the eggs are then ingested with the foliage. The human warble fly, *Cordylobia anthropophaga,* lays its eggs in sand tainted with urine; the larvae remain inactive until stimulated by the vibrations and rise in temperature associated with the visit to the site by another host. In the latter case, the urine probably provides an adequate chemical stimulus; however, in those parasites that must locate the plant habitat of their host, the clues used are poorly known; visual, chemical, tactile, and even sound stimuli are all possibilities. Laboratory experiments conducted by Cade (1975) indicate that the tachid fly, *Euphasiopteryx ochracea,* locates its cricket hosts by orienting to the latter's song. Both parasites and predators of scolytids have been shown to be attracted to the odors (terpenes) given off by the bark beetle's host trees as well as the pheromones released by the beetles while the trees are under attack. Ullyett (1953) discovered that ichneumonid wasps, *Pimpla bicolor,* were attracted from a distance to a ruptured cocoon of its moth host.

Once a host has been located, female parasites often tap it a number of times with their antennae indicating that perhaps tactile and contact chemical stimuli are involved in host recognition. The ovipositor of the hymenopterous parasites are also richly endowed with sensory receptors, apparently used to locate hosts feeding beneath a substrate, and perhaps in host recognition.

BEHAVIORAL MECHANISM OF FOOD ACQUISITION

In spite of extensive research the results of which appear in an enormous body of entomological literature, our understanding of food finding and recognition by insects is far from complete. However, regardless of the basic food type utilized, host selection is clearly a catenary process. The suggestion of Dethier (1954), that the behavior of phytophagous insects consists of three component phases, seems equally applicable to blood feeders, parasites, and others. Dethier's phases are (1) orientation to the food, (2) initial biting response, and (3) continued feeding. Thornsteinson (1960) recommends that dispersal be added as a fourth phase that would complete a regenerating cycle, but whether or not that is valid depends on one's basic concept of what constitutes dispersal (see Chapter 4). The host selection behavior of entomophagous insects has been divided into phases defined as (1) host habitat finding, (2) host finding, (3) host acceptance, and (4) host suitability (DeBach, 1964), but the first two phases appear to be a subdivision of the first phase of Dethier while the remaining two phases in each case seem parallel.

Each of the phases defined by Dethier is in turn made up of a sequence of responses, often to stimuli of various kinds (visual, chemical, tactile), that cannot be broken; that is, each response is dependent upon completion of the one that precedes it as in the response chains of Wigglesworth (1953). One must realize, however, that each response in the chain is not necessarily a reaction to a single isolated stimulus. Stimuli obviously interact, as when the convection currents generated by a warm-blooded organism carry olfactory cues perceived by biting flies. Furthermore, one kind of stimulus may prime an organism to respond to a different kind of stimulus as when a butterfly is primed to land on a piece of colored paper when the scent of a flower is added to the air (Tinbergen, 1951).

One of the difficulties that has impeded our progress in gaining an understanding of host selection has been the inability to observe those links in the chain that result in orientation to the host from a distance. Host finding can involve two basic kinds of behavior:

1. Nondirected behavior (such as klinokineses) which results in a change in the frequency of random turns so that the probability of host contact is increased, and orthokineses, which arrest random search.
2. Directed behavior (taxes), which results in orientation to a specific host stimulus or group of stimuli.

There seems to have been a preoccupation with host attraction and orientation to a source of stimulation, particularly chemical stimulation. Most laboratory experiments dealing with these subjects have been conducted in restrictive arenas and with olfactometers that employ odor-carrying airstreams that do not allow the insects to maneuver normally. There is, however, sufficient evidence amassed from field observations to suggest that both directed and nondirected behavior is often involved. Insects with a variety of food relationships, as well as those that use pheromones for reproductive assembly, demonstrate what can best be described as an odor-released or odor-induced amenotaxis (response to an air current). This certainly seems to be the case in the mate-finding behavior of males of the noctuid moth, *Trichoplusia ni* (Shorey, 1964), and probably fits the early stages of host searching by mosquitoes (Laarman, 1955), some phytophagous species (e.g., McMullen and Atkins, 1962), and probably carrion feeders (Haskell, 1966). Since the air is seldom still and even a small amount of turbulence caused by convection curents and objects in the environment would result in odor plumes rather than uninterrupted gradients (see Chapman, 1967), klinokineses can be expected to play an important role in host finding. The searching insect would simply fly upwind on a more or less straight course, as long as it was adequately stimulated by the appropriate chemical; however, once the scent was lost, it would begin to fly back and forth until the odor was detected again.

The sequence of responses that results in actual acceptance of the host and sustained feeding is more readily observable. At close range, visual, tactile, and contact chemical stimuli may all play a role in maintaining contact between an insect and its host, the initiation of probing or exploratory feeding, and finally sustained feeding. The termination of feeding if the insect remains undisturbed occurs in response to proprioreceptors associated with the digestive tract. The final stimulus that determines whether an insect will feed or not is most commonly of a chemical nature. Nutrients, frequently sugars, are strong feeding stimulants, but some of the odd nonnutritive substances discussed earlier are important to the feeding behavior of certain species. For example, the cabbage aphid, *Brevicoryne brassicae,* requires sinigrin, common in the mustard family, as a feeding stimulus; when it is present, the aphid will feed on abnormal hosts.

The following partial description of feeding behavior in the blow fly, *Phormia regina,* by Dethier (1966) is probably typical of the feeding response sequences of many insects: "Stimulation of chemoreceptors on the tarsi triggers proboscis extension; extension places the chemosensory aboral hairs of the labellum in contact with the food; stimulation of the hairs results in spreading of the labellar lobes which places oral taste papillae in contact with the food; stimulation of the papillae triggers and drives ingestion . . . "

Finally, one must bear in mind that the feeding behavior of any insect is regulated not only by its own endogenous rhythms and developmental physiology, but also by the life cycle and activity rhythm of its host species. Furthermore, most species function in relation to the constraints imposed by environmental factors such as photoperiod, temperature, and so forth. Hocking (1971) presented a diagram showing the interaction of cyclical changes that influence the feeding behavior of blood feeders (Figure 8–2), but a similar scheme should apply to any species regardless of its basic trophic relationship.

Figure 8–2. Interacting cyclical changes that influence the attraction of blood-sucking insects to their hosts. Similar constraints regulate the feeding behavior of herbivores and entomophagous species as well. (From B. Hocking, 1971, *Ann. Rev. Entomol.,* **16:** 1–26.)

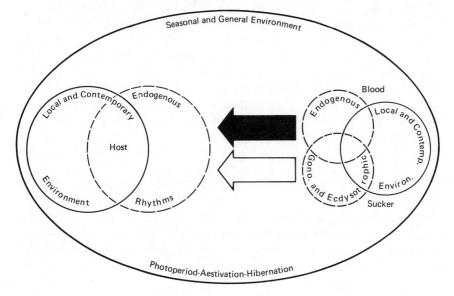

CHAPTER 9

Defense

The subject of defensive behavior among insects is a complex one encompassing structural, physiological, biochemical, and behavioral adaptations. However, a consideration of defense belongs in a book such as this because virtually all of the morphological and biochemical mechanisms that have developed have been accompanied by the evolution of appropriate patterns of behavior that make them effective. For instance, the protective coloration of an insect would be of little value if the animal's behavior did not bring it to rest on an appropriately colored background.

At least to people unfamiliar with their ways, insects often seem to be more on the offensive than the defensive, especially when the observer comes under attack. There is no doubt that some insects, like blackflies and horseflies, are so aggressive in their attempts to acquire a blood meal that their hosts would view them as offensive; the same seems true of attacks by bees or wasps attempting to defend a nest that has been disturbed. However, neither of these displays of aggressive behavior constitute offense, and the latter example is clearly one of defense. In fact, among the insects there are relatively few conflicts among members of the same species, and those that occur between species are usually of a predator-prey nature. In those species such as dragonflies, crickets, and sphecid wasps, in which the males defend territories or battle over mates, the fighting is more ritualistic than injurious (see Chapter 7).

Self-defense against parasites and predators is a basic biological problem faced by all species. It is a particularly significant problem for insects because their success as a group stems to a considerable extent from their great mobility—a factor that exposes them to attack by a variety of other organisms. Consequently, the defensive strategies that have been evolved by insects are both diverse and remarkable. The fantastic attention to detail that nature has paid in the evolution of defensive traits provides ample evidence of both the intensity and the variety of selective pressures that the insects have been subjected to. The major mechanisms employed in self-defense can be classified mainly as behavioral, structural, chemical, and colorational. However, such a classification is justifiable on

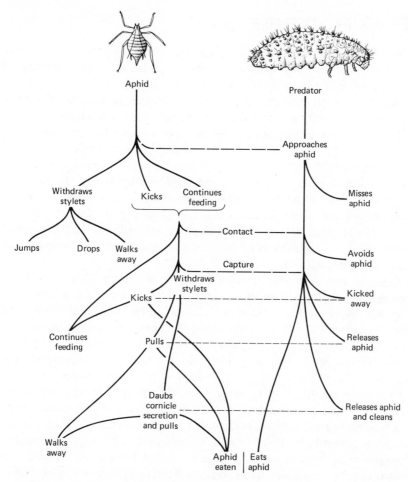

Figure 9–1. The possible interactions resulting from an encounter between a predator and its aphid prey. (Redrawn from A. F. G. Dixon, 1973, *Biology of Aphids,* Edward Arnold & Co., London.)

no other basis than that it provides an organized framework for discussion. Although there are some examples of purely behavioral defense that involve no other components, almost all defensive strategies have a behavioral component. Furthermore, many insects employ more than one defensive tactic, either in sequence or in combination, and there is often a great deal of behavioral integration involved (Figure 9–1); I hope that the following organization of this subject will not mask these characteristics. Systemic defensive mechanisms such as phagocytosis, encapsulation, immunity, and so forth, will not be considered; readers interested in these areas are referred to Salt. (1970).

BEHAVIORAL DEFENSE

As already indicated, behavior is an important component of all forms of defense, but there are some insects that rely almost entirely on behavior for protection. Perhaps the most straightforward approach to defense is to flee; this is extremely effective among the insects because of their small size and their ability to accelerate very rapidly. Many leafhoppers (Cicadellidae) move rapidly sideways to the opposite side of the branch they are on and simply attempt to stay out of sight. Taking flight is a particularly effective means of escape, as anyone who has attempted to capture a fly or a grasshopper with bare hands will readily testify. In fact, the selective pressure of predation almost certainly played a major role in the evolution of the insect flight mechanism. Jumping is also an effective means of escape in both flying and flightless individuals. Wingless aphids successfully avoid predation by larval coccinellid beetles by jumping suddenly (Wratten, 1976).

Reflex dropping is also an effective form of escape. Behavior of this type is displayed by a number of kinds of beetles, especially weevils, and a variety of caterpillars. Geometrid larvae often drop from their food plant on a strand of silk when disturbed and then reel themselves back up when the danger has passed. Other caterpillars, like those of the California tussock moth, just drop freely to the underlying vegetation. The opposite of rapid movement is, obviously, to remain motionless. Remaining motionless reduces visibility or gives the appearance of being inanimate. Numerous insects adopt a posture that feigns death (**thanatosis**) as a defense against natural enemies that prefer live prey. A death posture functions particularly effectively in combination with rapid movement. If an adversary is persistent, the surprise change from a lifeless pose to a sudden evasive movement introduces an element of surprise, which effectively confuses the pursuer.

Another form of mainly behavioral defense involves the adoption of a threatening posture, as in the katydid *Neobarettia spinosa* (Figure 9–2). Such poses can be quite effective in frightening adversaries sufficiently so that they will at least cease to be aggressive. Threatening postures are often enhanced by bizarre structural features, which make the insect appear grotesque, or by patches of bright color, which when displayed suddenly change an attacker's search image. Some threatening postures actually mimic the defensive behavior of other species that use an additional mechanism as their first line of defense. This can be seen among nonstinging species that engage in the abdominal thrusting behavior characteristic of stinging species. One of the most fascinating threatening postures is adopted by the larvae of the Brazilian sphingid, *Leucorrhampha*, which, at rest, blend well with their normal background. When disturbed, the larva twists its body to expose a pattern of scalelike mark-

Figure 9–2. The katydid *Neobarettia spinosa* in its threatening posture. (Reproduced with permission of T. J. Cohn.)

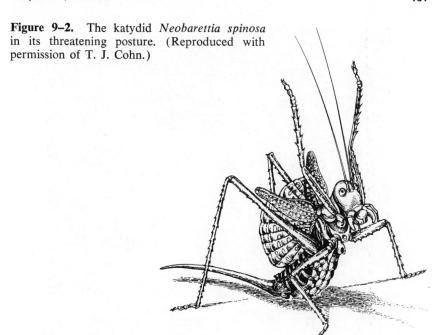

ings and then sways back and forth in the manner of a small snake (Figure 9–3).

A special kind of behavioral defense involves the construction of an individual protective shelter or the utilization of a wide variety of natural crevices and holes formed as a by-product of feeding. Boring insects, for example, create a protective place in which to live as they tunnel through the woody parts of plants. Others, such as the ambrosia beetles, excavate tunnels beneath the bark as a place in which to culture their fungus food and rear their young. Many species use plant parts or other materials to construct protective cases in which to rest or pupate. In the broadest sense, a case can be considered as any extra covering intentionally formed by an insect, regardless of whether it is composed of a body secretion or largely of foreign material. Thus the frothy secretion of spittle bugs (Figure 9–4), the wax threads of woolly aphids, and the resinous shells of scale insects can, along with silken cocoons, be considered as extraorganismic protective devices.

The larval lepidopterans are the predominant group among the terrestrial case makers. Some fasten their cases to fixed objects; others construct portable cases that they carry about. Often the case of the last larval instar is sealed to serve as a pupation site. The coleophorids construct tubelike cases of silk, leaf material, and feces that cover their entire body with the exception of the head capsule. When feeding, the larva is attached to the plant by its mouth parts so that its case stands out

Figure 9–3. The caterpillar of *Leucorrhampha ornata* assuming its defensive posture in which it resembles a small snake. (Courtesy of C. W. Rettenmeyer.)

from the plant surface. The bagworms (Psychidae) live in portable cases and move about freely, carrying their abodes wherever they go; when it is time to pupate, the case becomes a pupation site. Their cases are frequently made from twigs or leaf pieces bound together with silk in a species-specific fashion (Figure 9–5).

Among the aquatic insects, the caddisflies are by far the best-known case builders. As in the related Lepidoptera, trichopterans construct both permanent and portable cases from a wide range of materials, including bits of leaves, twigs, sand grains, and pebbles that are bound together with silk into a tubular structure. When larval development is complete, the case is firmly secured and sealed for pupation. For many insects the

Figure 9–4. Typical frothy protective secretion produced by a spittle bug, Cercopidae: Homoptera. (Photograph by M. D. Atkins.)

case is so distinctive that it can be used for purposes of identification (Figure 9–6).

Many of the protective cases constructed by insects probably evolved primarily in response to the selective pressures of the physical environ-

Figure 9–5. Examples of the portable protective cases constructed by the larvae of psychid moths known as bagworms. (Redrawn from F. Nanninga, 1970, in *Insects of Australia,* Melbourne University Press.)

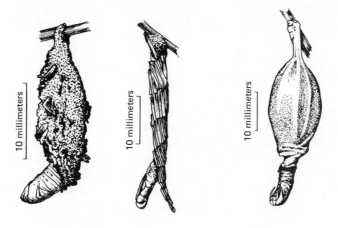

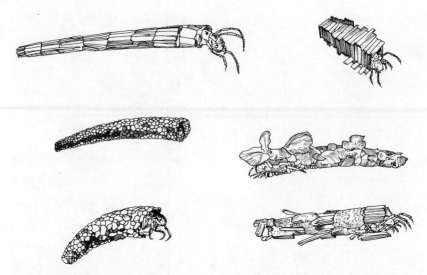

Figure 9–6. Cases constructed of a variety of different materials by the larvae of caddisflies. (Redrawn from Quick, 1970, in *Insects of Australia,* Melbourne University Press.)

ment, although they certainly provide some protection against natural enemies. However, many parasites have evolved countermeasures, such as long ovipositions, that have reduced the effectiveness of some protective structures.

Some insects gain protection simply by piling debris on their backs. The larvae of the datura beetle (Chrysomelidae) carry about accumulations of their own sticky fecal material, which contains a high concentration of plant alkaloids that most predators find unpalatable (Figure 9–7). The masked hunter (Reduviidae) carries a load of lint and dust stuck to its dorsal surface as an effective means of camouflage. The predaceous larvae of some brown lacewings (Hemerobiidae) are called trash carriers because they pile the remains of their victims upon their backs.

STRUCTURAL DEFENSE

In some insects, the integument becomes so heavily sclerotized that it provides adequate protection against almost any form of natural attack, including the beaks of insectivorous birds. This is especially true of beetles, among which the ironclad beetles (Tenebrionidae) and some weevils (Curculionidae) are well known by collectors because of the difficulty of penetrating them with mounting pins. In some ants, the front of the well-sclerotized head capsule of the soldier caste is flattened and

Figure 9–7. Larva of the datura beetle with a protective dorsal shield consisting of its own excrement. (Photograph by M. D. Atkins.)

can be used to plug temporary openings in the nest against intruders (Figure 9–8).

The mouth parts of insects are particularly useful in defense, as are the modified cerci that form the forceps at the tip of the abdomen of earwigs; the latter become proficient at picking up adversaries, such as ants, and throwing them to one side with a flick of the abdomen. Insects do, however, have a variety of specialized defensive structures, mainly in the form of spurs and modified setae. The leg spines of insects are primarily an aid to locomotion but become effective weapons when the legs are raked across a victim. Caterpillars belonging to several families have a cloak of hairlike setae that make them unpalatable to a variety of insectivores and increase the difficulty of oviposition upon them by parasites. Short fragments of these setae break off and readily enter the skin, causing irritation and festering similar to that caused by a bark sliver. There is no question as to the effectiveness of these hairs as a protection against predators. During outbreaks of the western tent caterpillar, *Malacosoma pluviale,* for example, birds of all kinds completely ignore what would appear to be an abundance of food.

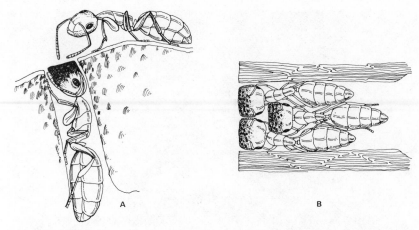

Figure 9–8. Nest-guarding behavior of soldiers of the European ant *Camponotus truncatus.* (A) A lone soldier blocking a small opening. (B) A group of soldiers blocking a large entrance hole. (Redrawn from E. O. Wilson, 1971, after Szabó-Patay, *The Insect Societies,* The Belknap Press of Harvard University Press, Cambridge, Mass.)

CHEMICAL DEFENSE

Arthropods in general and insects in particular display the widest diversity of chemical defenses of any group of terrestrial animals. Chemical defense among the insects can be divided broadly into the use of compounds referred to as **venoms**, injected by a skin-perforating apparatus, and the use of odoriferous or repugnatorial substances sequestered from their food or produced by special integumentary glands. Thus the chemicals used for defense can be either manufactured within the insect (**endogenous**) or obtained from some outside source (**exogenous**). Arthropod venoms were reviewed by Beard (1963), whereas other forms of chemical defense were reviewed by Roth and Eisner (1962) and Eisner (1970).

For the most part, endogenous defensive chemicals are produced by multicellular structures comprised of glandular epithelium and a saclike reservoir with a cuticular lining in which the secretion is stored. The defensive compound is usually discharged by muscular or hydrostatic compression of the reservoir or by its evagination. Several staphylinid beetles have tubular glands that, under pressure of the hemolymph, turn inside out like fingers of a glove (Figure 9–9).

Nonvenomous defensive secretions are highly variable. Roth and Eisner (1962) and Eisner and Meinwald (1966) listed numerous defensive compounds that are used singly or in combination by arthropods. Not all such compounds are insect-made. Often they are simply special chemical

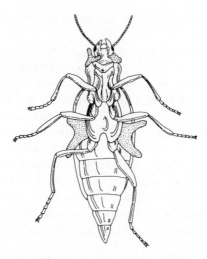

Figure 9–9. Ventral view of the staphylinid beetle, *Malachius bipustulatus,* showing evertible defensive glands. (Redrawn from R. Jeannel, 1960, *Introduction to Entomology,* trans. by Harold Oldroyd, Hutchinson Publishing Group, London.)

constituents of the diet that, when sequestered by an insect, render it unpalatable. Probably the best-known example is the sequestration of the cardiac glycocides of milkweed plants by the larvae and adults of the monarch butterfly. Other insects, lepidopterans in particular, are rendered undesirable as prey by various other plant-derived compounds that become incorporated into their tissues. Obviously, the potential predators of such insects have no way of knowing that these prey are unpalatable unless they try them and learn that they should be avoided. Apparently, the learning of predators is enhanced if distastefulness is associated with a specific visual image. In the course of these predator-prey relationships, many unpalatable species have evolved patterns of bright-colored markings, often referred to as **warning** or **aposematic coloration**. This will be discussed briefly under colorational defense later in this chapter.

In some species, the distasteful components of an insect's food are actually concentrated for release against an adversary before it takes a trial taste. The larvae of some swallowtail butterflies have a thoracic pouchlike structure called an **osmeterium**, where odiferous plant components accumulate and from which they are expelled when the pouch is everted in response to a disturbance. One well-known example is the larva of the anis swallowtail, which upon disturbance everts a bright orange-red osmeterium and its contents of volatile anis oil. Recently, Roth and Eisner discovered that a man-made herbicide known to be repellent to ants is sequestered by a species of grasshopper when it feeds on sprayed foliage;

the herbicide is incorporated into the froth the grasshoppers release for defense.

Defensive chemicals that are not injected can be separated broadly as being odoriferous or repugnatorial. Numerous insects, including roaches, earwigs, several families of beetles, and several families of bugs, have rather characteristic odors that our own sense organs readily detect as undesirable. Many of these compounds are not released merely in the presence of danger but are released constantly to the surface of the integument from the epithelial glands that produce them. Anyone who has handled the large black "stink beetles" (Tenebrionidae) or whirligig beetles (Gyrinidae) knows how long these odors can linger. Indeed, they may cause an entire insect collection to smell for months after the specimens have been pinned and dried. One species of ant produces **citral**, a compound closely related to citronella, commonly used in commercial insect repellents.

Most of the chemicals used in insect defense function largely as repellents that deter attackers, but in higher concentrations some of these substances are highly toxic. Some chemicals, however, seem to be primarily toxic. The meloids are often referred to as blister beetles because a compound called **cantharidin** in their integument is a powerful mucous membrane irritant and vesicant, which was once thought to be useful as a human aphrodisiac. Another compound, **pederin**, produced by staphylinid beetles of the genus *Paederus,* produces a dermatitis. A **saponin** compound that is extracted from a beetle and used by Kalahari bushmen to coat their arrows is a powerful paralytic. The formicine ants derive their name from the fact that they are stingless and rely on **formic acid** secretions in their attack on other organisms and in colony defense. The formic acid is applied topically as a spray that these ants can propel up to 30 cm, or subcutaneously through wounds caused by their mandibles.

Perhaps the most often described example of chemical defense encountered among the insects is that of the famous bombardier beetles of the carabid genus *Brachinus*. These small black beetles derive their name from the audible report that accompanies the explosive emission of their defensive secretion. By raising and rotating the tip of its abdomen, a bombardier beetle can accurately aim its "cannon" at an adversary. The internal defensive apparatus consists of a reservoir into which hydroquinones and hydrogen peroxide are secreted by associated glands. The two compounds pass posteriorly into a cuticular chamber where they come in contact with a catalytic enzyme (Figure 9–10). The reaction that occurs when the three compounds are mixed together results in a sudden liberation of oxygen that expels a visible cloud of quinone at the enemy (Eisner and Meinwald, 1966).

In the broad sense, venoms include all forms of toxic compounds that are injected. This definition would therefore encompass the chemicals

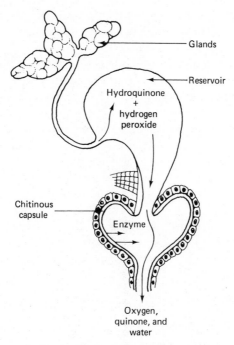

Figure 9–10. Defensive system of the bombardier beetle *Brachinus.*
(Adapted from H. Schildknecht and K. Holoubek, 1961, *Angew. Chem.,*
73 (1): 1–6.)

produced by epithelial cells associated with setae, urticating hairs, and
spines; salivary secretions injected by the mouth parts; plus the venoms
of aculeate hymenopterans. Unlike nonvenomous compounds, venoms are
usually ineffectual when applied topically. However, when injected into
an organism, venoms induce their characteristic response. Some venoms,
such as the salivary secretions of many predaceous insects, have a general
toxicity because of their primary function as aids to digestion. These
tend to degrade or liquify the surrounding tissue in an irreversible manner.
In some blood-feeding species, such as mosquitoes, these secretions contain an anticoagulant that facilitates feeding by retarding clotting.

The venoms produced by the sting glands of aculeate hymenopterans
may have evolved from the paralyzing venoms employed by the parasitic
wasps. Some paralyzing venoms have a fairly widespread effect, but others
are specific to only a few host species. Such specificity appears to be
attributable to the chemical nature of the compounds rather than to the
host preference of the parasite. Furthermore, many venoms are unstable
and produce only a temporary paralysis that accommodates oviposition,
whereas others produce permanent paralysis. The venoms of social bees,
wasps, and ants are used mainly for the defense of the colony. The

chemical nature of insect venoms is highly variable, and only a few have been adequately studied from a biochemical standpoint.

The social insects often enhance their defensive capability through coordinated group behavior. Angered bees and wasps will often respond to danger with a rush of defenders from the nest; group stinging behavior may then result in response to an alarm pheromone released in association with the venom. An alarm pheromone has also been found to occur among several species of aphid. When attacked by a predator, the prey aphid produces a droplet from its cornicles. A volatile component of the secretion repels other aphids within a distance of 1 to 3 centimeters (Nault, Edwards, and Styer, 1973). Ants also employ alarm pheromones that draw workers to a disturbance in the vicinity of the nest (Wilson, 1963). Other ants will encircle a portion of food and discharge a defensive secretion outward to repel the foragers of other species. Workers of the weaver ant, *Oecophylla longinoda,* mark, and in a sense defend, their territories against invasions by alien workers (Hölldobler and Wilson, 1977).

In some termite species, specialized soldiers known as nasuti (Figure 9–11) eject a sticky frontal gland secretion from a projection of the head that resembles a nozzle. When workers are traveling back and

Figure 9–11. Scanning electron micrograph of the head of a termite nasute, *Nasutitermes exitiosus,* viewed from below. Note the droplet of defensive secretion on the tip of the snout. (Courtesy of Thomas Eisner, Cornell University.)

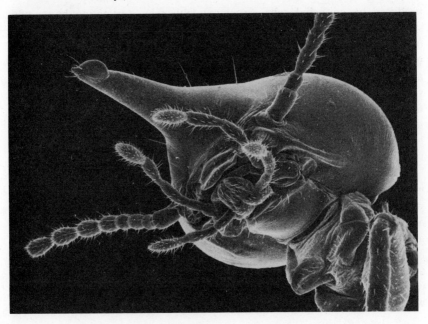

forth over exposed trails, the nasuti line up along the sides of the pathways with their snouts pointing outward to provide protection to the foragers. A group of nasuti in a cooperative effort will also encircle an adversary and discharge the secretion, which hardens into a sticky thread. By swinging their heads from side to side, they can throw loops into the threads and quickly entangle an enemy much larger than themselves.

COLORATIONAL DEFENSE

Defensive or, more correctly, protective color patterns can be grouped under four categories: cryptic coloration, flash patterns, warning signals, and mimicry.

Cryptic coloration

Cryptic coloration is simply camouflage. Crypsis can result from either blending into a featureless background or looking like a particular object that forms a common component of the environment. The success of this form of protection is highly dependent on the closeness of the match between the color and pattern of a species and that of its habitat background, but it also depends on the existence of appropriate behavioral mechanisms. A cryptic individual must select a suitable background on which to rest and must either remain motionless or move in an appropriate fashion, such as the gentle swaying motion displayed by some leaflike species. For example, the mottled color pattern of the grasshopper genus *Philippiaeris,* from the rocky desert areas of central Chile, ranges from a light sandy beige to a dark green. I collected a number of these interesting creatures (Figure 9–12) and invariably found them to rest among stones that matched their color almost perfectly.

Certainly, one of the truly fascinating aspects of protective coloration in general, and of cryptic patterns in particular, is the level of perfection that has evolved in so many species, as some of the following examples will demonstrate. The study of industrial melanism discussed in Chapter 12 indicates the survival value of camouflage and how quickly changes in a color pattern can occur when driven by the selective pressure of predation. The long evolutionary history of insects has provided ample opportunity for a high degree of refinement in their protective color patterns.

Protective coloration also provides numerous examples of similar patterns having evolved in totally unrelated groups that occupy similar habitats where they are subjected to the same kind of selective pressures. The structural similarity between caterpillars and the larvae of sawflies is quite remarkable, but the similarity in the color patterns of these groups is truly amazing. Members of both groups, which feed on the needles

Figure 9–12. The Chilean desert grasshopper of the genus *Philippiacris* camouflaged among the small stones that typify its habitat. (Photograph by M. D. Atkins.)

of coniferous trees, are often the green color of the needles and bear similar longitudinal, light-green stripes that resemble the midrib and characteristic lighting patterns of these elongate leaves.

Several basic color patterns are employed that help insects to blend well into featureless backgrounds. Disruptive coloration tends to break up the outline of the body, with light and dark markings that suit a mottled background or create the impression of a random pattern of sunlight and shadow (Figure 9–13). No matter how well an organism's color matches its background, the insect will reveal its position readily if it casts a shadow. Shadow elimination by compressing the body and wings tightly against the substrate therefore becomes an important behavioral aspect of camouflage. Even when an organism is colored to match its background, predators with stereoscopic vision can detect the presence of a prey as the result of the three-dimensional image created by differential lighting. This is overcome quite well by patterns of countershading. A solid-colored caterpillar, for instance, would be highly visible on a leaf of similar color because of the higher level of light reflectance from its dorsal surface compared to that of its lateral surfaces. If an individual is more darkly colored along the middorsal line and there is a gradation to

Figure 9–13. The disruptive coloration of the wings of a brushfooted butterfly. (Courtesy of Richard Elzinga, Kansas State University.)

lighter color along the sides, it has the effect of flattening out the visual image perceived by a predator.

Only a few insects are capable of adaptive color change such as that displayed by chameleons and a few fishes. Some locusts can vary their color from nearly white through yellow and brown to almost black and thereby match the color of their backgrounds. The change is not as rapid as in some vertebrates but comes about gradually, supposedly in response to the light intensity reflected from the background. Similarly, the color patterns developed by the pupae of certain lepidopterans are determined in advance by the background color perceived in the larval stage (Wigglesworth, 1964). Several genera of stick insects, however, display a chromatic rhythm, being paler by day than by night. This change is the result of the movement of pigment granules in the epithelial cells, which is under the control of hormones.

Not all insects live in habitats with a generally featureless background into which they can blend. In habitats characterized by open space, where an individual might frequently form a silhouette against a rather uniform color field such as the sky, better protection is afforded by resemblance to a commonly occurring object. This similarity to inanimate objects is usually referred to as **homomorphism** (shape), **homochromism** (color), or **homotypism** (form and color), whereas mimicry is more properly reserved for the resemblance to another animal.

A common kind of homotypism displayed by insects involves the resemblance to leaves, twigs, and floral parts. Leaf look-alikes, including katydids, stick insects, and butterflies, present a wide variety of examples that illustrate the incredible perfection evolution can produce. In this regard, one of the interesting aspects of leaf resemblance emerges in the examination of a series of species with activity periods that span the passage of the seasons. In accordance with the changing condition of the leaves among which they hide, the wing veins as well as markings on the wing surface of these insects simulate the appropriate pattern of leaf venation and configuration. Insects that are active during the spring, when the leaves are fresh and unmarred, are green and look like perfect new leaves. Those that are active later in the year may resemble tattered older leaves, discolored and damaged through the course of time. These patterns are achieved with irregular wing margins and clear areas that resemble perforations, combined with blotches of color that resemble physical damage or splashes of bird droppings (Figure 9–14).

Larval stages also resemble inanimate objects. The caterpillars of some sphingid moths have countershading and diagonal markings, which make the individual look like a rolled-up leaf, complete with a stalk provided by a thick terminal spine. Geometrid moth larvae, or loopers, often resemble twigs, complete with budlike swellings and appropriate bark markings (Figure 9–15).

Figure 9–14. The protective leaflike appearance of the forewings of a katydid. (Courtesy of C. W. Rettenmeyer.)

Figure 9–15.   The protective twiglike appearance of a geometrid cater-pillar. (Courtesy of C. W. Rettenmeyer.)

One cannot examine the cryptic coloration and behavior of insects without wondering how it is possible for such perfect copies of noninsect objects to have evolved and how much protection they afford the species that have evolved them. If they do not occur by chance, why are all species not protected in this manner? Crypsis is just one of a number of defensive strategies, and it is best suited for those species that are sedentary, diurnal, and live in well-lighted habitats. Furthermore, as perfect as camouflage may be, it can never provide complete protection. For example, some birds pick up certain kinds of objects at random, and, as long as they encounter a reasonable number of prey items for their effort, they are likely to continue and even concentrate on those objects that their prey resembles most closely (de Ruiter, 1952).

Obviously, if the prey becomes more difficult to see, a predacious species must improve its search ability in order to survive. If we recognize a coevolution between the hunter and the hunted, the members of the latter group that most perfectly blend into their surroundings will always have the best chance to survive predation and will consequently leave propor-tionately more offspring than the variants that are more easily seen. There is a sizable body of experimental evidence that illustrates the protec-tive value of a slight resemblance to a noninsect object, even under cage conditions where the search effort of a predator is greatly reduced. Since some apparently poor crypsis reduces predation at least a little, we can see how there would be enough selection to drive evolution toward better and better crypsis. Thus, after countless generations, we see an accumulation of inherited characteristics, which in some cases border on cryptic perfection. The fact that some of the details cannot be seen clearly without the aid of a magnifying glass (an objection sometimes raised by skeptics) is a concern that stems from an anthropocentric view of the natural world.

Flash patterns

Even the best camouflage does not fool all of the predators all of the time, so many species have evolved a second line of colorational defense, referred to as flash patterns. The basis of the effectiveness of flash patterns is to induce a rapid change of the search image of the pursuing species. This is accomplished by quickly changing from cryptic to conspicuous and back again. The efficacy of the flash pattern strategy is amply demonstrated by grasshoppers with brightly colored hind wings. When resting among the vegetation, many grasshoppers are well camouflaged but may be found with a little careful searching. As you approach close enough to make a capture, your quarry will often take flight, during which the colorful hind wings are exposed. This changes your search image, and, by the time you have made a mental adjustment, the grasshopper drops out of sight among the vegetation, once again relying on its cryptic coloration for concealment.

Some flash patterns are apparently more effective than others. Concentric light and dark rings, usually referred to as **eye spots**, seem to be particularly effective. In some insects, eye spots are displayed perhaps as a means of drawing attention away from some vital part such as the head, but they seem more effective in deterring the attack of a predator when displayed suddenly. A number of lepidopterans, such as the eyed hawk moths, are cryptically colored in their normal resting position, but, when disturbed, they expose a pair of prominent eye spots on their hind wings (Figure 9–16). Eye spots may actually serve a dual purpose. They are certainly an effective means of altering a predator's search image but they may induce a fright response as well. Blest (1957) showed that birds, which had not had any learning experience with such markings, regularly drew away from peacock butterflies (Nymphalidae) that suddenly displayed their eye spots. The fact that the birds in Blest's experiment drew away suggests that they were at least startled by the sudden appearance of an ominous pair of eyes. Blest also showed that the birds in his experiment withdrew when other shapes, such as a pair of crosses or a pair of single rings, were exposed suddenly. This suggests that eye spots could have evolved in a series of steps from more simple markings.

Figure 9–16. The defensive display of eye spots. (A) In the normal resting position the moth relies on camouflage; (B) when disturbed, the forewings are spread displaying eyelike markings on the hind wings. (Courtesy of C. W. Rettenmeyer.)

Warning coloration

Whereas some insects are perfectly camouflaged, others seem to flaunt themselves with gaudy displays of color and ornamentation. Most brightly colored insects are adequately protected by special defensive structures, chemicals, or by the fact they are unpalatable. However, this is not always the case, as illustrated by those species that resemble flowers as a form of cryptic coloration. The extremely showy color patterns of many foul-tasting butterflies and beetles provide an obvious association between the color pattern and distastefulness that is readily learned by vertebrate predators while they are young. With periodic reinforcement of the learned association, experienced predators generally avoid prey displaying similar patterns. Likewise, the alternating bands of colors presented by the abdomens of stinging hymenopterans provide a specific image that can be remembered by a predator that has been stung. One can demonstrate the effectiveness of warning coloration quite simply. If an inexperienced frog is fed a bumblebee from which the sting has been removed, it will subsequently take an intact bee. However, once having been stung, the frog will refuse subsequent offerings of similar appearance.

Mimicry

The concept of mimicry was first proposed more than 100 years ago by the English naturalist Henry W. Bates. On the basis of observations he made on an expedition to the Amazon basin, Bates proposed that species with overlapping ranges (**sympatric**) could evolve a resemblance if the common color pattern was beneficial to both species. Bates had observed that some palatable species had similar color patterns to sympatric unpalatable species. He concluded that the unpalatable species (the **models**) were employing warning coloration, and the palatable species (the **mimics**) derive benefit from the fact that they are not attacked by predators that have learned the association between the color and bad taste.

The preceding concept is known as **Batesian mimicry** and has been the subject of controversy ever since it was proposed. It seems to rule out the possibility that species come to resemble each other as a result of parallelism. However, it does not seem necessary to consider every case of resemblance as a case of mimicry. The predominance of black and red patterns in the wings of South American butterflies compared to a predominance of black and green patterns among African and black and blue patterns among Indian species is well known. We might expect, therefore, to encounter more similarity in the appearance of groups of species within each of these geographic regions than between them.

Opponents of the concept of mimicry argue that the gradual accumulation of traits that seems to function suitably in the development of crypsis could not have operated effectively in the development of mimicry. Although a mimic survives because it looks like a model that predators have learned to reject, it is questionable whether a species could gain sufficient protection from early minor changes for such changes to be selected. We must remember, however, that many mimetic patterns function while the species are in motion, so perhaps they do not need to be as highly refined as patterns employed in crypsis; camouflage must withstand the careful search of a predator, often very close at hand. It can be counter-argued that such changes occurred as the result of major mutations or had their beginning in convergence. No matter which side of the argument one chooses to take, it is a matter of fact that there are many pairs of species that look alike in which one is well defended and the other is defenseless. There are also a number of species that retain their original color pattern in some parts of their range but in other areas are clearly similar in appearance to unpalatable species with which they overlap. Finally, there seem to be just too many examples of mimicry in which details of the resemblance are extraordinary and result in some obvious benefit to the mimic to pass it all off as an accident of parallelism or convergence.

For Batesian mimicry to remain effective, it is obvious that there must be a numerical balance between the model and the mimic such that a predator does not have more experiences with palatable than with un-palatable prey of the same appearance. For a mimetic pattern to evolve, a palatable species must be considerably less abundant than the model, and it must remain so in order to benefit from the similarity. This raises an interesting question as to how the development of a mimetic pattern by a palatable species affects the level of predation on the unpalatable species. Certainly, the learning of young predators would be slowed by encounters with palatable prey. This could increase the predation pressure on the model and lead to the selection of divergent color patterns. Viewed in this way, the gradual development of mimicry seems completely possible, as both model and mimic could evolve together rather than only the mimic having to undergo a large-scale change.

The naturalist Müller noticed that in some groups of mimetic species all of the individuals are unpalatable. Consequently, when two or more unpalatable sympatric species have a similar appearance, we call it **Müllerian mimicry**. At least in theory, it would seem to be beneficial for a group of species to adopt a common warning signal rather than different signals. A predator would have to learn to recognize only one pattern instead of several separately. Assuming that the predator requires a number of bad experiences to learn the appearance of an inedible prey, a group of inedible species could share the number of individuals that must be sacrificed to educate the predator.

Various workers have argued that closely related species may look

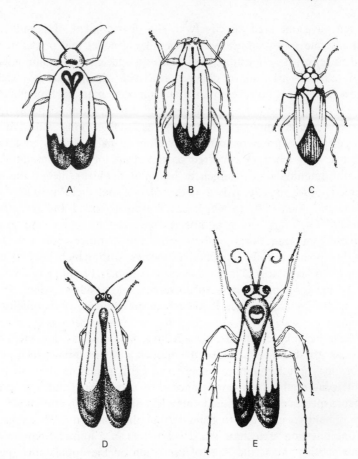

Figure 9–17. A mimicry-ring consisting of (A) the distasteful soft-shelled beetle *Lycus rostratus,* (B) a cerambycid beetle, (C) a lygaeid bug, (D) a butterfly, and (E) a pompilid wasp. (Redrawn from *Mimicry* by Wolfgang Wickler. Copyright © W. Wickler 1968. Used with permission of McGraw-Hill Book Company.)

alike because of a common ancestry. For example, two genera of black and yellow striped wasps probably inherited the pattern from a similarly colored ancestor. In such a case, we cannot consider one to be the model and the other the mimic, since, in a strict sense, there is no mimicry at all. However, this technicality does not seem as important as the more fundamental question as to whether Müllerian mimicry is effective, which it apparently is. For example, in South America there is a group of Müllerian mimics consisting of four different families. In addition to the Müllerian species, there are two species of edible Batesian mimics, and all of them are avoided by predators (Wickler, 1968).

Mimicry is commonly thought of as similarity in color patterns because of the influence of the early workers who mainly studied butterflies.

However, it is obvious that color similarity alone would not be as effective as similarity in over-all appearance; there are in fact many examples of insects from different orders that resemble each other in both form and color. One of the most diverse groups that forms a so-called mimicry ring involves a cerambycid beetle, a hemipteran, a butterfly, and a spider-wasp; all bear a rather striking similarity to a highly unpalatable soft-shelled beetle (Figure 9–17).

There are also many examples of mimicry that rely mainly on form in that both model and mimic are monochromic. In some cases, the benefit that befalls the mimic is obvious, as in the case of the defenseless long-horn beetles (Cerambycidae) that closely resemble foul-smelling tenebrionids (Figure 9–18). In other cases, as with many ant mimics, the significance of the resemblance is obscure. In spite of their chemical defense, ants are eaten by birds, reptiles, amphibians, and mammals, so it seems unlikely that ant mimics would gain much protection against predation simply through resemblance. It may help some species to mix with ants and thereby remain unnoticed, even to the extent of becoming permanent residents in their nests, but there are many successful ant guests that do not look like their hosts. Perhaps the answer lies in the fact that because ants are generally numerous, a few individuals of another species could move among them without being conspicuous and thereby enjoy an improved chance of survival. Whatever the basis for it, the similarity between ants and totally unrelated insects, such as the small hemipteran illustrated in Figure 9–19, can be truly remarkable.

Figure 9–18. A defenseless, monochromatic ceramycid beetle (left) that mimics a foul-smelling tenebrionid beetle (right) with which it is sympatric. (Photograph by M. D. Atkins.)

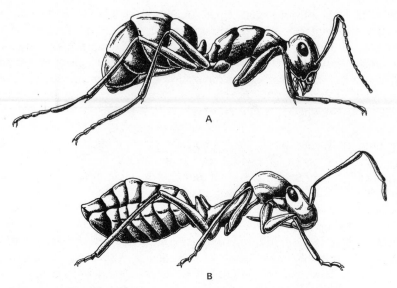

Figure 9–19. An example of ant mimicry. (A) Small black ant from southern California and (B) a small, ground-living, myrid bug found in the same area. Note the elbowed antennae, constricted waist, and pedicle-like wings of the bug that improve its antlike appearance.

GROUP DEFENSE

Group defense is well known as an important behavioral adaptation of eusocial insects. Termite workers are summoned to block openings and repair the damage to their nest, and nasuti function in groups to encircle an attacker and entangle it in strands of sticky frontal gland secretion. Bees, wasps, and ants are all known for their group stinging behavior in defense of their nests, and the specialized soldier caste of ants acts in groups to defend columns of foraging workers.

However, group defensive behavior is not restricted to eusocial species. Good experimental evidence for the effectiveness of such behavior among the more generalized hymenopterans known as sawflies has been published (Prop, 1960; Tostowaryk, 1972; Knerer and Atwood, 1973). Careful study of other gregarious and colonial species may reveal that group defense is actually quite common.

Larval sawflies of several species of *Neodiprion* and *Diprion* exhibit group defensive displays comprised of synchronous jerking motions of their anterior and posterior ends, and may also extrude a sticky resinous material from their mouthparts. Prop (1960) found that this behavior discouraged both parasites and avian predators. Tostowaryk (1972) found that when *Neodiprion* larvae were dispersed, the pentatomid bug, *Podisus modestus,* could attack an individual larva and retreat with it without disturbing neighboring larvae. However, when the sawfly larvae were in a compact group, the struggle associated with the predator's attack of an individual stimulated a group defensive response by the remaining members of the colony. Even if the predator's attack was successful, it became smeared with the larval exudate, which slowed subsequent attacks. The end results of the group defensive behavior were an increase in the number of unsuccessful attacks by *P. modestus* and fewer attack attempts per unit of time.

CHAPTER 10

Parental care
and presocial behavior

Very few insects spend their lives in total isolation. Most at least come together to mate sometime in the course of their life cycle, and various means of communication have evolved that enhance this behavior (see Chapter 7). Nevertheless, insects display a behavioral continuum that ranges from a strong solitary instinct to one that is truly social (see Chapter 13). In many species, solitary and gregarious behavior are not mutually exclusive. Some caterpillars, for instance, are highly gregarious during their early instars and seem able to move about in an orderly fashion only when a certain individual assumes the lead (Figure 10–1); yet by their last larval instar they have become solitary.

Most forms of nonsolitary behavior would seem to be associated with some aspect of procreation. Numerous species swarm or engage in other kinds of more or less gregarious behavior associated with mate location and pairing. Once copulation has been completed, the pair usually, but not always, splits up and the female goes about her egg laying in solitude. Frequently, the female lays her eggs and never sees the young she has produced; she may exercise little care in placing her eggs, as is the case among some tropical butterflies that broadcast their eggs over the forest canopy, or a great deal of care, as exemplified by some parasitic wasps (see Chapter 8). However, the females of species scattered throughout most of the insect orders engage in some degree of parental care, and in a relatively few species the male cooperates with the female in the raising of the young.

The simplest form of parental care is displayed by the females of species that guard their eggs or young during their early life. Even this minimal amount of protection clearly has a selective advantage over the abandonment of the eggs immediately after they have been laid, and is a form of behavior engaged in by both solitary and gregarious species. Several of the true bugs, particularly the pentatomids, lay their eggs in the open and then shelter them with their body. The European earwig, *Forficula auricularia,* is gregarious throughout much of the year. In the fall the young adults pair and pass the winter as a couple in a chamber in the soil. In the spring the mated female drives out her mate and enlarges the earthen chamber to accommodate a cluster of up to about 100 eggs, which she lays. She then tends her eggs to keep them free of mold and wards off potential predators with her forceps. After the eggs hatch, she continues to care for her young through their first two nymphal stages.

The web spinners (Embioptera) also care for their eggs and young, but, unlike the earwigs, embiopterans remain gregarious, each group sharing a system of interconnecting silken tunnels. The eggs are attached to the inner surface of the tunnels and are often camouflaged with plant debris. After the young hatch, they add to the network of tunnels and live with the parents in a communal group.

More elaborate parental care and some cooperation between males and females occur among the beetles. The bess beetle, *Popilius disjunctus* (Passalidae), forms poorly organized colonies in decaying logs and stumps. The adults and larvae communicate with each other by stridulating, and the adults, apparently of both sexes, masticate wood fragments and mix them with saliva to form a preparation they feed to their young.

Dung beetles (Scarabaeidae) show various degrees of complexity in the behavior associated with the care of their young that seem to suggest an evolutionary progression. In its simplest form, the brood care consists of the female fashioning a ball of dung that is placed in a shallow burrow; the female subsequently oviposits on the dungball, covers it with dirt, and moves off to excavate another burrow elsewhere. In other species, the male assists the female to form the dungball and may guard it while the female prepares the hole in which it will be buried; this represents a simple division of labor. In the genus *Copris,* the male and female co-operate in the excavation of the nest, plus the gathering and preparation of food for the larvae. In one species, the female guards and cleans the nest throughout the period of larval development. In a related species, both parents remain throughout the life of the larvae, but only the female cleans and maintains the nest (Wilson, 1971).

Fairly advanced levels of presocial behavior are displayed by the bark and ambrosia beetles (Scolytidae). In the Douglas-fir beetle, *Dendroctonus pseudotsugae,* the female initiates a brood gallery by chewing through the outer bark of the host tree to the interface between the wood and the inner bark. Males and other females are attracted to the host in response to a pheromone released by the attacking female (see Chapter 6). Each female is joined by a male that initially sits in the gallery entrance and produces a chirping sound. The purpose of the chirping sound is not clearly under-stood but may serve to communicate the fact that pairing has occurred. The female then proceeds to construct a gallery oriented with the grain of the wood and lays small groups of eggs in niches cut in the side walls of the tunnel. As the main gallery is lengthened, the male keeps it clear of

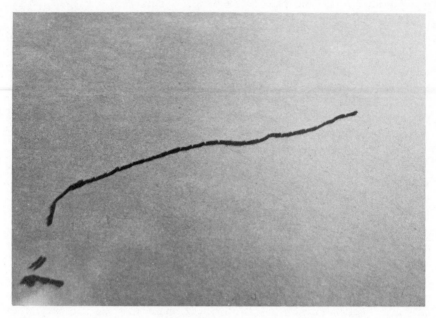

A

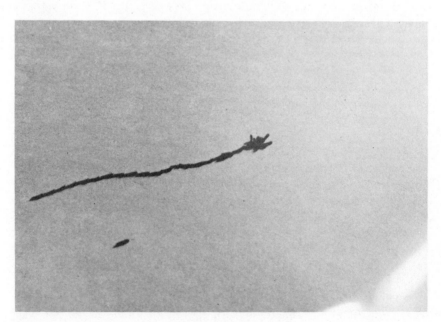

B

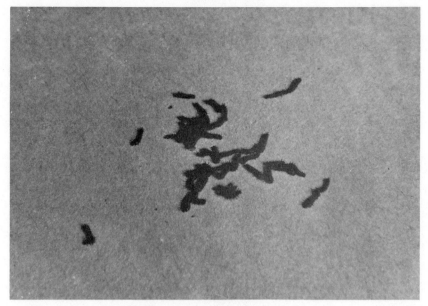

C

Figure 10–1. Processionary behavior of first instar *Hemileuca electra* larvae. (A) A normal procession. (B) Confusion following the removal of the leader (below the column) causing a cessation of forward progress. (C) Complete disorientation of the group a few minutes later. (Photographs by M. D. Atkins.)

boring dust and excrement by forcing debris from the entrance with his legs and body. Eventually, the gallery opening is sealed with a plug of boring dust that presumably prevents the entrance of natural enemies. Once the entrance is sealed, the male cuts one or more ventilation holes through the bark. The larvae mine away from the main gallery (Figure 10–2), so there is little opportunity for contact between the female and her offspring even though she often remains in the gallery throughout larval development.

The level of presocial organization in the ambrosia beetles is still further advanced. In the genus *Gnathotrichus,* it is the male that initiates the brood gallery and releases the aggregating pheromone. Once joined by a female, both parents cooperate in excavation of the main gallery, which is innoculated with fungal spores carried by the male in special structures called **mycangia**. The female lays eggs singly in cradles carved into the side walls of the main gallery and provides fungus for the larvae. The developing larvae feed on fungus that grows on the wood chips and the feces they produce while enlarging their niches. The larvae do not mine away from the main gallery as the bark beetles do but are located in

such a way that they can be contacted and given food by their mother. The female remains throughout larval development, during which time she tends her brood and keeps the gallery system free of debris.

According to Wilson (1971), the most advanced form of parental care known among the Coleoptera is displayed by the burying beetle, genus *Necrophorus* (Silphidae). Male and female cooperate in the excavation of an earthen burrow and in the preparation of the ball of carrion that is interred within. Once the larval food has been lowered into the burrow, the entrance is sealed from within. The eggs are laid in a depression in the carrion; but, when the larvae hatch, they not only feed on the putrifying flesh, but also receive a brown liquid passed directly to their mouths from the mouth of their mother. If the larvae are deprived of the food provided by their mother, they are unable to complete their development. In two other species, the male also participates in feeding the brood (Wilson, 1971). The social behavior of these beetles is also more advanced than most in that they display some degree of **altruism**. Small members of the same species may assist in the internment of the carcass and then depart without taking any further part in reproduction. The dominant pair then take over the food supply (Milne and Milne, 1976).

PRESOCIAL BEHAVIOR OF HYMENOPTERA

No group of insects displays more diversity of parental care and social behavior than the Hymenoptera. The wasps, bees, and ants all have truly social (**eusocial**) representatives, but most members of the former two groups are presocial. Both the bees and the wasps display an ascending hierarchy of behavior that many investigators believe represents an evolutionary sequence similar to that which led to the development of eusociality. Evans (1958) presented one such schema for the wasps, based on both morphological and behavioral adaptations. A simplification of Evans' behavioral sequence is presented in Table 10–1.

The aculeate (stinging) Hymenoptera are believed to have evolved from parasitic forms, and the behavior of many solitary, predatory wasps is clearly related to that of their parasitic ancestors, except that the ovipositor is used to narcotize their prey. In the more primitive species, the female searches for an appropriate prey, paralyzes it with her sting, lays an egg on its surface, and departs. These prey, left unprotected, are

---

Figure 10–2. The brood gallery of the Douglas-fir beetle, *Dendroctonus pseudotsugae,* showing the larval mines progressing away from the central adult gallery. Once the larvae begin to feed, they no longer have contact with their parents. (Courtesy of the Pacific Forest Research Centre, Victoria, B.C.)

Table 10–1. Major Steps in Development of Eusocial Behavior of Wasps (The families that demonstrate each level of behavior are given in parentheses.)

Step of Behavioral Sequence	Comments
1. Prey–egg (Pompilidae)	The female locates a prey, temporarily paralyzes it with her sting, lays eggs on the prey, and departs; the prey recovers and carries the wasp larvae which feed as external parasitoids
2. Prey–natural crevice–egg (Pompilidae, Sphecidae)	The female drags the paralyzed prey to an available protective crevice where it is left with an egg attached; the female thus provides young with a level of protection
3. Prey–nest–egg (Pompilidae, Sphecidae)	Female paralyzes a prey and then constructs a nest in which it is placed along with an egg; this is a slightly advanced level of parental care
4. Nest–prey–egg (Pompilidae, Sphecidae)	Same as step 3 except the nest is constructed before the prey is captured; this introduces homing in that the female must return to a previously selected nest site
5. Nest–prey–egg–prey (Sphecidae)	Similar to step 4, but the addition of more prey after egg is laid introduces mass provisioning as a more advanced form of parental care
6. Nest–prey–egg–prey–prey (Sphecidae, Eumenidae)	Instead of mass provisioning, the nest is provisioned progressively with fresh prey; this brings the female into contact with her developing offspring; in some species, the female remains in the nest when not provisioning, thereby reducing predation, and may also clean the nest of partially consumed food
7. Prey macerated by female (Eumenidae, Vespidae)	In the process of progressive provisioning, the fresh prey are macerated by the female and fed to the larvae; this brings the female into direct contact with her offspring and provides opportunity for trophallaxis and the transfer of pheromones
8. Female life prolonged and offspring remain with nest (Vespidae)	The prolonged female life results in overlap with the first generation offspring, which remain and lay eggs in cells they add to the nest; this results in small colonies consisting of the mother and a group of undifferentiated daughters
9. Trophallaxis and division of labor (Vespidae)	Mother and daughters cooperate in nest building and the care of young, but there is no permanent division of worker and

Table 10–1. *continued*

Step of Behavioral Sequence	Comments
	egg-laying castes; trophallaxis paves the way for queen dominance
10. Queen dominance (Vespidae)	The original offspring are all females that are incapable of producing their own female offspring, thus separating the reproductive and worker castes; intermediates may be common
11. Differential larval feeding (Vespidae)	Differential feeding of the larvae and trophallaxis lead to the production of a well-defined worker caste strongly differentiated from the queen, and a reduction in the number of intermediates.

Source: After H. E. Evans, The evolution of social life in wasps. *Proc. 10th Int. Congr. Entomol, Montreal,* **2**:449–457, 1956.

clearly subject to consumption by other organisms or to accidental destruction before the wasp larva can complete its development.

The tendency of other species to conceal their prey, either before or after they have deposited their egg on it, is a form of parental care that has obvious survival benefits. Placing a single prey in a burrow in the manner of some *Ammophila* species is the full extent of parental care displayed by a number of wasps, yet it requires a great deal of effort (see Chapter 2). These species utilize single prey such as caterpillars that are large enough to fulfill the nutritional needs of their young. The progressive provisioning displayed by the sphecids, for example, not only permits the utilization of smaller prey that can be carried back to the nest in flight, but also has the potential for improving larval nutrition. The repeated visits of the female to the nest also provides an opportunity for her to come in contact with her young and opens the way to the exchange of hormones, pheromones, and gut symbionts, along with partially digested food (**trophallaxis**), and the possibility for parental control over the development of offspring, as seen in the burying beetle *Necrophorus*.

The development of patterns of behavior that increased provisioning seems to have been accompanied by the construction of fairly elaborate nests. Among the wasps, the simplest nests consist of no more than burrows in the soil. However, the preparation, provisioning, and final concealment of these burrows involves a behavioral sequence that is truly fascinating to observe. More elaborate nests are fashioned from a variety of natural materials and often are quite characteristic of the constructing species (Figure 10–3). A number of wasp species use mud for nest construction, making many trips to fetch enough to build a nest suitable for a single offspring. The time and energy devoted to the construction

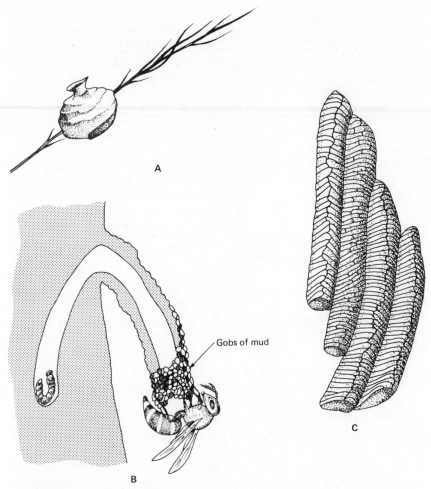

Figure 10–3. Some examples of nests constructed by solitary hymenopterans. (A) Juglike nest of the potter wasp *Eumenes*. (B) Section through the nest of the solitary wasp *Oplomerus* showing the storage tunnel provisioned with a caterpillar and the down-curved entrance tube constructed of mud. (C) Group of mud nest units of the pipe organ mud dauber *Trypoxylon*.

of these elaborate nests might seem to be an excessive form of parental care; however, the resulting improved survival of the young has clearly reduced the number of progeny each female needs to produce to assure survival of the species and thus permits some species to specialize on less-abundant food sources.

Bees, like wasps, display many degrees of presocial behavior. Within small groups of related species it is possible to identify numerous steps in the evolutionary progression from solitary to eusocial behavior. As mentioned earlier, a major difference between the bees and the wasps

Figure 10–4. A close-up photograph of the soil surface of a nesting area of the alkali bee, *Nomia melanderi,* showing the density of the entrance holes. (Courtesy of William Stephen, Oregon State University.)

lies in their diets and the morphological adaptations related to food collection. Adult bees not only feed on nectar and honey but raise their young on pollen or a pollen-honey mixture, rather than on macerated insects. Because both honey and nectar are readily storable, the larval cells of bees can be provisioned initially rather than progressively throughout development, as is the case with many wasps; this tends to reduce the occurrence of contact between adult bees and their developing offspring.

In the solitary bees, each female independently constructs one or more nests containing a group of larval cells, which are mass-provisioned with sufficient pollen and nectar to fulfill the full nutritional requirements of the larvae. Such is the way of the leaf-cutter bee, *Megachile rotundata,* which nests above ground in preexisting holes of appropriate diameter and depth. The female, after locating a suitable nest site, constructs a series of larval cells in the tubular hole by lining the tunnel with pieces cut from leaves and petals. Each cell is provisioned with 7 to 12 loads of pollen and then separated from the next by several discs of leaf material. In her life span of about 6 weeks, each *M. rotundata* female will construct 5 or 6 such nests, totaling approximately 30 larval cells (Stephen, 1973).

Other solitary bees, like the soil-nesting alkali bee, *Nomia melanderi,* form nesting aggregations. The female constructs a more or less vertical burrow and, like the leaf-cutter bee, constructs a series of larval cells mass provisioned with pollen. The suitability of certain patches of soil often results in the construction of a number of nests in one area (Figure 10–4), but there is no cooperation between individual females.

Perhaps because of the tendency to mass provision made possible by their food type, the bees do not show the variety of parental care behavior displayed by the wasps. However, species of *Nomia* construct composite nests consisting of a communal entrance with cells off on side branches that are constructed and provisioned by individual females. Michener (1969) classifies these species as communal and suggests that they benefit from an economy of nest construction labor. The regular back-and-forth movement of a group of females using a single entrance probably enhances the defense of the brood as well.

The parental care and brood-rearing behavior of some presocial wasps and bees is clearly only slightly less elaborate than that of eusocial species. However, the formation of large social groups, involving a division of labor and queen dominance, requires a higher level of behavioral coordination as we will see in Chapter 13.

PART III

SOME SPECIAL CONSIDERATIONS

The foregoing chapters were concerned primarily with the behavior of individuals, or of species interpreted from the "average" behavior of individuals. It was of course necessary to realize that intraspecific and interspecific interactions contributed to the development of these species specific behavioral patterns. Nevertheless, these influences were kept in the background in order to emphasize the way that insects meet their biological needs behaviorally. In the first three of the four remaining chapters more emphasis will be placed on the importance of the inter-actions between individuals.

Chapter 11 concentrates on the variability of the behavior encountered among members of a population rather than the average picture of behavior assembled from a number of individual elements. This change in viewpoint is important to our understanding of the great adaptability of the insects; it also has important implications in population ecology and management.

Chapter 12 presents the genetic aspects of behavior as a basis for an appreciation of the fact that behavior and structure evolve according to the same mechanisms and are often interdependent. It also describes the way that behavior can contribute to speciation in overlapping populations and thereby explain, at least impart, the great diversity of the insects.

No book on insect behavior would be complete without at least a brief review of truly social (eusocial) behavior; Chapter 13 presents such an overview. Not only does social behavior among insects reveal the ultimate in what can be achieved through the integration of inherited responses, it provides us with some insight into our own social interactions. Furthermore, the strongest evidence for learned behavior among the insects has come from the study of eusocial species.

Chapter 14 departs from the theme that ties the previous thirteen chapters together in that it presents a more pragmatic view. Nevertheless, it provides a mechanism for synthesizing various behavioral elements that may come to bear on a subject of applied significance. Furthermore, it points out the weakness in some common approaches to entomological problems that have arisen from our failure to recognize the importance of insect behavior.

CHAPTER 11

Population behavior

Wellington (1957) opened one of his most significant papers with the sentence: "Populations are composed of individuals and individuals differ." At the time, Wellington was concerned about the failure of population theorists to recognize that populations are composed of behaviorally, physiologically, and genetically different groups, the relative proportions of which may change as the population density changes. The oversight that Wellington identified was rather remarkable in view of the fact that modern evolutionary theory is based on the concept that the variability inherent in populations is the raw material upon which selective pressures act. Although Wellington's initial paper and several subsequent ones (Wellington, 1959; 1960a; 1964; 1977) were well received, the importance of the qualitative differences of individuals within a population still takes a back seat in studies of insect population dynamics to the more apparent direct causes of numerical change. Individual differences of various kinds can be important to population change, but none of these differences is more apparent than in behavior. Often insects that look identical and originate from the same parents behave quite differently. In fact, behavioral inconsistencies are frequently a source of frustration for entomologists.

Wellington's investigation of the population cycles of the western tent caterpillar, *Malacosoma pluviale,* revealed rather remarkable differences in the behavior of both larvae and adults that contributed importantly to changes in the size and distribution of a local population. Under field conditions, newly hatched larvae congregate on the egg mass from which they emerge. After a day or so the group constructs a communal tent of silk that serves as a shelter from which the larvae emerge to feed each day. This group behavior apparently was responsible for the fact that, prior to Wellington's study, observers had disregarded the behavior of individual larvae. But Wellington soon discovered that isolated young larvae behaved quite differently in ways that later proved to have major biological importance; furthermore, the basic behavioral differences displayed by young larvae were retained as they passed through subsequent instars and increased in size (Wellington, 1957).

It will not be possible to discuss all of the details of this now classic piece of research, but some of the findings that exemplify the importance of studying population behavior must be presented. Wellington separated first instar larvae on the basis of their ability to perform directed movements toward a single light source. Only a portion of the individuals were capable of moving toward the light independently, and these were termed **Type I larvae**. The rest of the larvae (termed **Type II**) displayed a range of behaviors that permitted the distinction between active and sluggish individuals. In the absence of a silk trail laid down by a Type I individual, Type II larvae tend to aggregate when in contact with each other or become inactive when isolated (Figure 11–1).

Each colony results from an egg mass laid by a single female; the

Figure 11–1. Contrasting behavior of Type I and Type II first instar larvae of *Malacosoma pluviale* placed on white paper in a beam from a single light source. (1–4) Type I larvae photographed at 0, 5, 10, and 20 seconds. (5–8) Type II larvae photographed at 0, 2, 9, and 10 minutes. (Drawn from photographs in W. G. Wellington, 1957, *Canadian Journal of Zoology,* **35:** 293–323, with permission of the author and the National Research Council of Canada.)

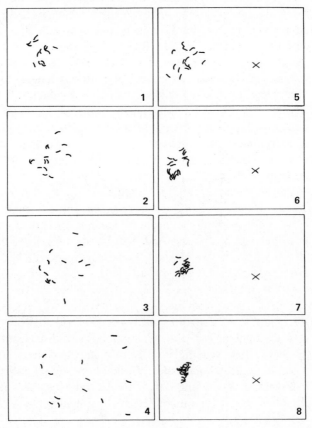

relative abundance of the various larval types varies from colony to colony. The differences in larval behavior and the make-up of a colony were found to affect feeding, tent construction (Figure 11–2), rate of development, and survival. Nevertheless, Wellington (1957) concluded that there is no ideal mix of Type I and the different subtypes of Type II larvae that would produce a single kind of colony effective at all population densities and in all environments within the habitat; depending on the circumstances, both Type I and active Type II larvae have potential beneficial or detrimental qualities. For example, Type I larvae orient precisely and respond quickly to change, thereby providing an important element of direction to the colony. Their high level of activity generally keeps the colony sufficiently stimulated to feed regularly and make frequent encounters with fresh foliage (food). The more mobile colonies are less subject to parasitism and less subject to an increase in virus infection. On the other hand, the tendency of Type I larvae to wander may increase the chance of parasite attack when parasites are abundant and, during periods of high colony density, may also increase the chance of contracting disease through contact with ailing individuals from other colonies.

In the case of this strongly colonial species, the superior Type I indi-

Figure 11–2. Different forms of tents constructed by fourth instar larvae of *Malacosoma pluviale*. The tent on the left was constructed by an active colony, the other by less-active colonies; the large tent fourth from the left was constructed by a colony that started out to be active but lost its better-quality individuals in the third instar. (Drawn from photographs in W. G. Wellington, 1964, *Can. Entomol.,* **96:** 436–451, with permission of the author and the Entomological Society of Canada.)

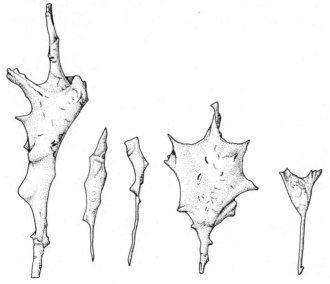

viduals are not only capable of surviving but also help to insulate the poorer-quality individuals from their own ineptitude. Type I larvae usually become active strong-flying adults, whereas the Type II larvae become less-active adults incapable of sustained flight. Consequently, there may be some emigration of the better stock, coupled with the local retention of the poorer stock. During favorable periods, this can result in the expansion of the population into outlying areas that are often environmentally marginal and where poor-quality colonies cannot survive. If the local environment permits the maximum survival of poor-quality individuals, the number of sluggish colonies will increase and further deteriorate in quality until they can no longer survive at all. The generally more favorable areas will then become repopulated by the emigration of better-quality moths from the peripheral areas (Wellington, 1965a; 1965b).

Although the western tent caterpillar is a colonial species, individual differences in behavior are known to occur among solitary species and, when studied more fully, will no doubt prove to be just as important. Ghent (1954), for example, found that the larvae of the jack pine sawfly, *Neodiprion banksianae* (which are gregarious in the first instar), begin feeding only after certain individuals take the first bite.

In my own work with the Douglas fir beetle, *Dendroctonus pseudotsugae,* I was unable to observe differences in larval behavior because to survive they must remain undisturbed beneath the bark. However, I did discover considerable behavioral variation in newly emerged adults and was later able to relate the differences to the developmental history of the larvae and their physiological condition (Atkins, 1966; 1967). The behavioral variability displayed by these beetles related to their responsiveness to hosts as well as to the sex and aggregating pheromones released by other individuals. Every sample of newly emerged individuals contained a portion that showed a strong tendency to migrate and a portion variously responsive to reproductive stimuli but less inclined to fly.

Later I was able to demonstrate that the behavior of the young adults was determined in the larval stage and was correlated with larval fat content. Large, heavy larvae, with a high fat content, became adults with a strong tendency to migrate, whereas poor-quality larvae become adults less inclined to fly, especially in the presence of reproductive stimuli. Unlike the tent caterpillar, however, adult Douglas fir beetles showed a change in behavior with time. That is, migrants would become responsive to host odors and pheromones after a period of flight and presumably the combustion of some stored fat (Atkins, 1969).

This variability in the behavior of a bark beetle population has important implications to the species' population dynamics similar to those discovered by Wellington for the tent caterpillar. Like *Malacosoma,* the Douglas fir beetle occupies a patchy environment, which can be exploited fully only if migrant adults spread out from population centers during favorable periods. In the tent caterpillar, those adults that do not emigrate contribute to the deterioration of the local population by increasing its proportion of poor-quality colonies. In the Douglas fir beetle, it is also possible for a local population to collapse due to a decline in quality but for a different reason. Instead of begetting more and more poor-quality colonies, nonmigrant Douglas fir beetles increase the density of a local population, which in turn increases the competition among developing larvae, thereby increasing the proportion of nonmigrants until there is drastic overpopulation. However, some migrants are always produced, which establish new population centers some distance away.

The two studies of population behavior discussed reveal the importance of intrinsic variability to the ecological fitness of the species and to their population dynamics. We will see in Chapter 14 how such information can be important to the development of sampling programs or the application of certain control strategies. Unfortunately, there are not many studies in which the significance of behavioral variability has been determined. Far too often, every attempt is made to eliminate variability, and this may account for the simplistic way in which many entomologists view behavior.

The failure to recognize the importance of individual differences within populations probably stems from the kinds of questions investigators ask and how they design experiments to answer those questions. Often the main objective is to determine what a species does under specific circumstances or how it reacts to specific stimuli. The experimental data contain a certain amount of variation, as one would expect, but this variation frequently does not find its way into the interpretation of the results for a variety of reasons.

For instance, the statistical treatments applied to the data often make variation less apparent. When an experiment is conducted with a specific goal in mind, the most common response or behavioral trait is accepted by the investigator as characterizing the species. Not only is emphasis placed upon the discussion of the most commonly displayed behavior, but degrees of response, angles of orientation, or other such measures may be averaged. This lumping tends to produce a description of behavior that parallels the "2.5 child" family.

Had Wellington chanced upon several poor-quality colonies of the western tent caterpillar when he initially collected larvae for his study and then accepted their disoriented behavior in his single-light experiment as being the norm, he never would have made his important discovery. Fortunately, he had a specific purpose in mind and formulated a hypothesis that required the examination of the total behavioral variation in his population. Thus, regardless of how infrequently a response occurred, its significance had to be examined. One cannot help but wonder how many important discoveries have been passed by because the investigator's objective did not require a certain kind of observation. The results of Wellington's work show the rewards of examining the entire array of "positive" and "negative" results when studying insect populations; "negative" results may not support the original hypothesis, but they may lead to new and perhaps more important ideas.

When one examines the techniques used in many behavior studies, it becomes clear that individual differences have been considered less important than average performance. The periodic observation of insects reacting to the conditions in an arena of some kind provides a good example. The end result, showing where a group of insects finally settles in a gradient of choice situation, may be much less important than what the members of the group did before they settled down; but these preliminary actions frequently go unrecorded (see Wellington, 1960b). As we have seen, the behavior of insects constitutes the integration of a varying number of responses to a variety of stimuli. How an individual responds in the presence of competing stimuli depends on the interplay of changeable response thresholds; the end result may ultimately be the same for two individuals even though some quite different responses may have been displayed beforehand.

Some recognition of the fact that specific end points are achieved in different ways and over different periods of time can be a revealing aspect of behavioral studies in terms of how the results might be used. It is therefore necessary to think of population behavior as a dynamic process that may be closely correlated with changes in abundance, sometimes as a cause, sometimes as an effect. Consequently, unless the objective is to analyze the genetic basis of a behavioral trait or simply to use an insect as a bioassay tool, there is little justification for developing a standardized laboratory population.

POPULATION BEHAVIOR AND ECOLOGY

Many modern ecologists view a species' niche as a multidimensional hypervolume characterized by the values of all of the resource parameters known to be essential to that particular species. This basic ecological concept places emphasis on the quality of the environment rather than the quality of the species or population (see Hutchinson, 1957; Macfadyen, 1957). Actually, this may be an unrealistic view. The individuals of a population are constantly responding to gradients of the resources that comprise their niche, and, as we have seen, behavior may be closely associated with population quality.

Competition for biotic resources, changes in the physical environment, and attack by natural enemies are dynamic, ongoing, selective pressures that tend to change the genetic make-up of a population. As was noted in the discussion of host selection (Chapter 8), changes in behavior caused by a change in a single gene provide the most rapid means by which an insect can adjust quickly to changes in its environment. The insects provide many examples of how changes in behavior have enabled closely related species to divide up key resources both spatially and temporally. As examples, Fitzgerald (1973) described three species of the bark-mining moth *Marmara* that had spatially divided the bark of green ash among them. Various wild bees are known to forage at different times of the day and thereby subdivide the nectar and pollen resources provided by a single species of flowering plant (e.g., Linsley, MacSwain, and Raven, 1963).

The preceding examples, and many others that could be cited, illustrate that each species evolves a strategy of efficiently utilizing the available time and energy to maximize its own survival. Some species, often referred to as **r strategists**, have evolved a way of life that maximizes their reproductive ability and their utilization of unstable environments. Other species, called **K strategists**, have evolved strategies that tend to maintain their populations at a more or less constant level, close to the carrying capacity of the environment. Usually, these different strategy groups have pronounced behavioral characteristics, and, in species with ranges that extend from stable environments (such as the humid tropics) into less stable ones (such as the temperate zone), notable behavioral differences characterize their extreme populations. The behavioral make-up of the population in an environmentally stable area would resemble that of a K strategist; the population at the other extreme of the range would behave more like an r strategist. In addition to this genetically based variation so important to the buffering of species populations against environmental change, there is also variability that results from the conditions of development (temperature, food quality and quantity, etc.) and behavioral differences related to physiological changes or conditioning.

In the laboratory it is possible to observe samples of a population of known age reared at a constant temperature on a standardized diet in the absence of competition. However, the behavior displayed by such a homogenous group may bear little relationship to that of a wild population. When I began to look for possible causes of behavioral variability in bark beetles, it became apparent that the environmental differences that might exist over a small distance could have a remarkable influence on behavior. Two of the most important factors proved to be the rate of larval development (related to temperature) and intraspecific competition. Rapid development and higher levels of competition resulted in smaller individuals with a lower total fat content. These beetles showed less inclination to migrate and a stronger response to host odors and pheromones than their larger and fatter siblings. Within a single log, the conditions for larval life can be so different that the emerging adults may display the extremes of behavior. For example, the attack density (number of family groups established) may be considerably higher on the exposed upper surface of a log than in the area resting on the ground. If the log is unshaded, the upper side experiences higher temperatures and greater diurnal temperature fluctuations. Consequently, the larvae on the upper side of the log develop more quickly and experience more competition for food than those perhaps only a foot away, on the bottom side of the log. Thus the adults that emerge from a small section of host material may vary markedly in their fat content and display strikingly different behavior.

If the environmental conditions could vary significantly enough to produce important differences in individual behavior, we might expect some difference in the behavioral composition of populations made up of members of the different generations of multivoltine species. A study of another bark beetle, *Ips paracenfusus,* supported this hypothesis (Hagen and Atkins, 1975). The population of adults that overwintered contained a significantly larger proportion of migrants than the adults resulting from the first spring generation.

Changes in behavior during the life of an insect have been recognized for some time (see Gross, 1913), but the change in the behavior of Douglas fir beetle adults added another dimension to the concept of behavioral variability, within a population synthesized by Wellington. Behavioral change could be induced by storing (aging) the beetles for several weeks under temperature conditions similar to those that sometimes occur in the spring and delay emergence of the young adults. But basic changes in behavior also occurred following several hours of flight exercise. These changes in behavior clearly have adaptive significance. A protracted preemergence period could have many implications for a population, including some reduction in numbers due to the mortality of weak individuals and losses associated with staggered emergence during brief spells of favorable weather; some differences in host physiology may also be encountered by late emerging individuals. Under such circumstances, the greater responsiveness (lower migratory tendency) resulting from the delayed emergence might act to buffer the population against further attrition due to migration and in-flight mortality. On the other hand, the behavioral changes that accompany flight are necessary for the establishment of new population loci.

Gaining an appreciation of behavior in general and behavioral variability in particular outside the less than natural confines of a laboratory arena is not an easy matter. Wellington chose an experimentally superb system in the western tent caterpillar. Its tent provided a relatively simple means of following numerical and distributional changes throughout the population as well as an indication of colony quality. Usually, we are fortunate if we can follow the behavior of a few individuals in the field for very long, and we cannot infer very much about what the total population is doing from such limited information. This has posed a problem in the interpretation of migratory behavior from local observations of a limited number of individuals (Johnson, 1969). The long-distance migratory flights of lepidopterans often involve time-space sequences that make any inference about navigation rather risky.

In spite of the difficulties and pitfalls of population behavior studies, they may yet provide the most important key to understanding insect population change. Insects seem to have a number of variable traits that affect their movement, survival, and reproductive capacity. Many of these traits are physiological and not easily identified without performing some test that would require the sacrifice of the individuals. However, just as morphological differences may express hidden genetic variation, so observable differences in behavior may reveal otherwise cryptic physiological variation. Behavior may then be used as a means of identifying individual differences that are more deep-seated and more important than the behavior itself. The ability to identify variants has substantial practical importance in the management of both beneficial and injurious species. The quality of the individuals that survive may prove to be a key factor in the determination of population trends.

CHAPTER 12

Behavior, genetics, evolution, and speciation

In the preceding chapters we have discussed a variety of behavioral adaptations that have contributed to the overall success of the insects. All of these behavioral traits have been inherited and selected for in the same manner as morphological traits. Behavior, however, adds another dimension to the evolutionary process because it represents not only the exploitation of structural features, but also a means by which certain environmental constraints can be lessened. For instance, many insects, particularly the eusocial species, employ stereotyped behavior patterns in the construction of nests that offer protection against natural enemies and provide an opportunity to regulate the physical environment.

Natural selection acts upon the inherited variability of every species; some variants are favored and increase, whereas others are eliminated. Often, the behavioral and morphological variability go hand in hand and may compliment each other. For example, the size and structure of wings of crickets influence the characteristics of their song. In the mole cricket genus *Gryllotalpa,* the males of *G. vineae* produce a louder and higher pitched song than those of *G. gryllotalpa.* However, the two species construct subterranean, horn-shaped burrows the physical properties of which are matched to the frequency of their respective songs (Bennett-Clark, 1976a, b). Thus the digging behavior enhances the characteristics of the song and presumably increases its perception by the females.

There are also examples in which morphological traits tend to compliment behavior. The phenomenon known as industrial melanism provides an excellent example. The British moth, *Biston betularia,* was originally described as a predominantly white moth with black speckles on the wings although some almost black individuals and various intermediately colored morphs of the same species were known to exist. During the day these moths rest on the lichen-encrusted trunks of native trees, but apparently the differently colored variants display no color discrimination in the selection of their resting places. During the latter part of the nineteenth century, industrial pollutants began killing the light-grey lichens and coating the tree bark with black deposits. At the same time, dark-colored moths became increasingly common around industrial centers, and by the early 1900s the populations in these localities consisted almost entirely of the dark (melanistic) form. Kettlewell (1961) reported the results of a study in which he released equal numbers of light and dark individuals in polluted areas adjacent to cities and in the pristine wooded countryside. He found that the survival of the light moths was greater away from the cities, whereas the survival of the dark moths was greater near the cities. The reason for this was that avian predators found and ate fewer dark moths when they were resting on the tree trunks darkened by soot and other pollutants. Conversely, the light moths were better camouflaged from predators in the country, where they blended into the background of resting places consisting of light bark and light-grey lichens. In areas where no industrial development occurred, the moth populations remained predominantly light colored. Furthermore, the reduced output of pollutants resulting from regulations imposed upon industrial concerns has allowed the light-colored lichens to grow back, and there has been an increase in the proportion of light-colored individuals in the populations of the moth in industrial areas.

Whereas morphological characteristics are normally fixed for an individual's life span (or at least for a particular stage of development), behavior may change as a result of learning, thereby adding a complication to the evolutionary formula. The capacity to learn is, of course, inherited but is generally of little importance as a behavioral modifier among insects. In fact, insects are often chosen for basic studies in behavioral genetics because their behavior is mainly stereotyped and their responses are unaltered by learning. This permits the investigator to measure a variety of basic responses and to quantify the effects of modification of the environment on their basic patterns of behavior.

Stereotyped, fixed action patterns can often be observed as independent elements of behavior in a controlled environment. This makes it possible to study the genes that govern behavior in the way one would study the genes that govern some structural trait. But, as with any genetic study, it is necessary to establish a basic unit of behavior with which one can work. Since behavior is so closely correlated with both morphology and physiology, it is probably determined by the operation of many genes. However, there are traits known to be governed by single genes that have provided considerable insight into the way behavior is determined.

SINGLE GENES AND BEHAVIOR

Single-gene traits are certainly the easiest to trace, regardless of whether they determine morphological, physiological, or behavioral traits. Those that produce a visible alteration in appearance are relatively simple to follow. Some genes affect more than one change (**pleiotropy**) and when one change is structural and the other is behavioral, it simplifies the analysis. However, it may be difficult to determine whether a gene produces only a behavioral change or a behavioral change accompanied by some nonapparent physiological change. Nevertheless, there are known single-gene effects on behavior that are both pure and pleiotropic.

The best-known example of single genes affecting behavior alone involves nest cleansing in the honeybee, *Apis mellifera,* which was carefully analyzed by Rothenbuler (1964). Larvae infected with American foulbrood eventually die, and their cadavers become a source of contamination that facilitates spread of the disease throughout the hive. Rothenbuler noted that bees differ in their response to the presence of diseased larvae; so-called hygienic strains removed foulbrood-killed larvae, whereas nonhygienic strains did not. This difference in behavior is determined by two independently segregating loci; one involves uncapping the cells containing diseased larvae and the other involves removal of the cadaver within. Hybrids between these two strains are all nonhygienic, indicating the recessive nature of the nest-cleansing behavior. Rothenbuhler backcrossed the hybrids to the hygienic strain and obtained groups that displayed the following four patterns of behavior in statistically equal proportions:

1. Bees uncapped cells containing dead larvae but left the cadavers in the cells.
2. Bees failed to uncap cells but removed cadavers from cells uncapped by the investigator.
3. Bees neither uncapped cells nor removed cadavers—**nonhygienic**.
4. Bees uncapped cells and disposed of cadavers—**hygienic**.

It is therefore possible to explain this behavior as being controlled by two pairs of alleles, one of which controls uncapping and one that controls the removal of cadavers. Since nonhygienic alleles are dominant, we can designate them *U* for uncapping and *R* for removal of cadavers; the alleles of the recessive hygienic strain would be designated *u* and *r*. The genetic constitution of the nonhygienic strain would be UURR and that of the hygienic strain uurr. The F_1 hybrids would be UuRr and would display nonhygienic behavior. A back-cross of the F_1 hybrid to the hygienic strain would produce the following:

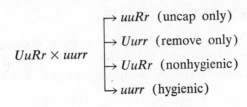

$$UuRr \times uurr$$

→ *uuRr* (uncap only)

→ *Uurr* (remove only)

→ *UuRr* (nonhygienic)

→ *uurr* (hygienic)

As far as is known, neither of the genes that determine hygienic or nonhygienic behavior cause either structural or physiological differences. However, most single genes known to influence behavior are pleiotropic. A number of mutant genes that produce conspicuous morphological changes in *Drosophila* have been shown to produce related changes in behavior. Only a few of the many known mutants that display significant differences in behavior from nonmutant strains will be presented as examples.

Mutations that affect eye characteristics are common in *Drosophila*. Since these changes affect vision, they often also affect patterns of behavior involving responses to visual stimuli. *Vermilion* and *bar* are two such mutant genes. *Drosophila* with the vermilion gene have bright red eyes compared to the normal, dull red eyes of wild flies. Studies (Bosiger, 1967) have shown that *vermilion* males mate more slowly and display lower mating success, presumably because they are less receptive to visual stimulation by the female. *Bar* mutants, which have narrower eyes with fewer facets, have difficulty locating females and receiving their visual signals during courtship. Both of these eye mutations are therefore a disadvantage in mating behavior. Another single-gene mutation known as *vestigial* results in males with shorter than normal wings. Since wing vibration is an important component of the courtship behavior of *Drosophila* and *vestigial* males cannot vibrate their wings in a manner that adequately stimulates the female, they exhibit poor mating success.

In Chapter 2 we examined the adaptive significance of circadian rhythms and noted that the synchronization of *Drosophila* pupal eclosion with the favorable temperature and moisture conditions that prevail at dawn was important to fly survival. Konopka and Benzer (1971) studied three single-gene mutations that altered the normal rhythm of *D. melanogaster*. One of the mutants is arrythmic and emerges throughout the day. Of the remaining two, one has a protracted cycle of 28 hours, and the other has a periodicity reduced to 19 hours.

CHROMOSOMES AND BEHAVIOR

In most sexually reproducing animals both the males and females are diploid. However, some insects, like male hymenopterans, are haploid. Cases of triploidy and tetraploidy are also known among insects, as are chromosomal mosaics (individuals having tissues with different chromosome complements). Changes in chromosomes can result from aberrations during cell division or from chromosome breakage, resulting in deletion, duplication, inversion, or translocation. However, inversions and changes in chromosome number are of most importance. Individuals with such abnormalities are excellent experimental animals for use in studies of the genetic basis of behavior.

Genetic polymorphism resulting from chromosome inversions is common among the flies and is of considerable interest because there must be a balance of selective pressures in order for individuals with different chromosomal constitution (**karyotypes**) to survive. Experiments have shown that the fitness of karyotypes differs, and inversion heterozygotes (heterokaryotypes) often display superior fitness to inversion homozygotes (homokaryotypes). Several studies have shown that the heterokaryote superiority in *Drosophila* is at least in part the result of more effective mating behavior, which is important to maintaining the polymorphism in natural populations. For example, Spiess and his coworkers noted large differences in the mating speed of homokaryotypes of *D. pseudoobscura*. Spiess and Langer (1964) noted that the frequency of occurrence of the chromosome inversions at the location where their stock was collected followed roughly the mating speed; AR and ST strains were most frequent, whereas the PP strain was least frequent (Figure 12–1). The more rapidly mating is completed, the sooner the eggs can be laid, and the sooner the larvae can become established and thereby get a headstart on later arrivals. Thus the fitness of the karyotype for rapid mating is increased because more surviving progeny (carriers of the genotype)

Figure 12–1. Cumulative percentage curve for matings between different strains of *Drosophila pseudoobscura* over a period of 1 hour. (Redrawn from E. Spiess and B. Langer, 1964, *Proc. Natl. Acad. Sci. USA*, **51:** 1015–1018.)

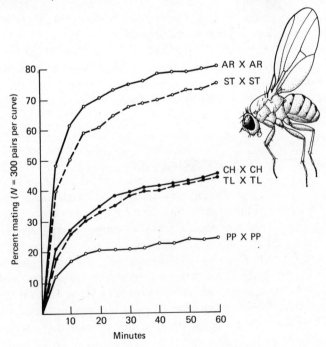

are contributed to subsequent generations than are carriers of other genotypes.

GENETIC MOSAICS

Genetic mosaics are organisms with more than one kind of tissue. **Gynandromorphs** are sexual genetic mosaics, which display a mixture of male and female tissue; there are also mosaics in which it is possible to distinguish between the effects of mutant genes and normal genes in one animal.

In most members of the Hymenoptera, males are produced from unfertilized eggs. Unmated females can only produce male offspring, whereas mated females can produce both fatherless sons and biparental daughters. Occasionally, eggs with two nuclei are produced, but only one nucleus becomes fertilized. The result is a sexual mosaic consisting of a mixture of male parts (haploid parts from the unfertilized nucleus) and female parts (diploid parts from the fertilized nucleus). Figure 12–2 illustrates a gynandromorph of the wasp *Habrobracon juglandis* along with a normal male and female.

Table 12–1 summarizing the behavior of 50 sexual mosaics of *H. juglandis* shows that the responses are usually normal for one or other sex rather than a mixture. The sex of the head seems to determine the nature of the behavior. Individuals with male heads were indifferent toward caterpillar hosts, and, even though a number of them had a typical female abdomen, they did not engage in the usual stinging thrusts and oviposition; these individuals did attempt to mount females. Those with

Figure 12–2. The parasitic wasp *Habrobracon juglandis*. (A) Normal female showing long wings, shorter antennae, and presence of a sting at the tip of the abdomen. (B) Normal male showing shorter wings, long antennae, and lack of sting. (C) Gynandromorph with malelike wings and antennae but sting of female. (Redrawn from P. W. Whiting, 1932, *J. Comp. Psychol.*, **14**: 345–363.)

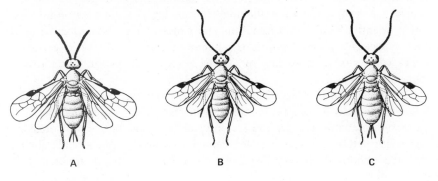

A   B   C

Table 12–1. Behavior of Gynandromorphs of the Wasp *Habrobracon juglandis* in Response to Females and Caterpillar Hosts

Predominate Sex of the			Reactions Toward Females		Reactions Toward Caterpillars	
Head	Abdomen	Number	Positive	Indifferent	Positive	Indifferent
Male	Mixed	9	9			9
	Female	20	20			15
Female	Male	1		1	1	
	Mixed	3		3	3	
Mixed	Male	2		2	2	
		1	1			
	Mixed	3	3			3
		3	3			
		1		1	1	
		1			1	
	Female	2	2			2
		1	1			
		3			3	
Total		50	39	7	11	29

Source: P. W. Whiting, Reproductive reactions of sex mosaics of a parasite wasp, *J. Comp. Psychol.,* **14**:345–363, 1932.

female heads were indifferent to females but reacted positively to caterpillars. Individuals with sexually mosaic heads displayed mixed and unpredictable reactions.

Sexual mosaics also occur in *Drosophila,* but, in addition, mosaic flies consisting of a mixture of mutant and normal genotype tissues have been produced in the laboratory. These flies have been used to produce two-dimensional embryonic fate maps that correlate abnormal behavior with the precise location of mutant structures and thereby gain a better understanding of the genetic basis of behavior. This technique has been applied by Benzer and his associates (see Benzer, 1973). In this way, investigators have shown that the leg-shaking behavior of a *Drosophila* mutant called **hyperkinetic** has its root in neurons of the thoracic ganglia of the central nervous system; the shaking of each leg is apparently controlled independently. Similarly, a mutation that causes some *Drosophila* to hold their wings in a vertical position, a behavior that would almost certainly affect wing vibrating aspects of courtship, has its base in the thoracic musculature.

How much can be learned about behavior from the analysis of mosaics remains to be seen. So far it has shown the strong nature of the link between structural changes and stereotyped behavior, but now it must be tested in relation to more complex behavior. Research has already shown that the abnormal ·rhythm displayed by some *Drosophila* apparently is associated with changes in the head (Konopka and Benzer, 1971). By applying the appropriate techniques, it may now be possible to identify those cells in the brain that control circadian rhythms.

MANY GENES AND BEHAVIOR

The relationships between genetics and behavior discussed thus far are simple in comparison to the complex control of behavioral traits by a number of separate genes.

Some traits vary according to an interplay of genotypic and phenotypic components. Selection experiments can be used to reduce phenotypic variability and then to localize the genes that control a quantitative trait to a specific chromosome. Most of the work of this type has again been done with *Drosophila,* chromosomes of which have been thoroughly mapped. A number of behavioral traits are under polygenic control in these flies; evidence for actual genetic loci exists for duration of copulation, geotaxis, and level of sexual isolation (Ehrman and Parsons, 1976).

Sexual isolation is the result of a number of mechanisms that have a behavioral component. Examples include seasonal and circadian rhythms and various aspects of courtship behavior that affect the attraction and acceptance of mates. The genetic basis of sexual isolation has been studied by following the distribution of a pair of chromosomes with the aid of mutant marker genes. One elaborate study of the sexual preferences of hybrids of a Central American and an Amazonian subspecies of *D. paulistorum* indicated that sexual isolation is fostered by genes on all of the three chromosomes possessed by the species. Furthermore, the effects were additive and, in concert, all but prevented crossing between the subspecies (Ehrman, 1961). In another study (Ehrman, 1960), F_1 hybrid females resulting from a cross between Amazonian and Andean-Brazilian subspecies of *D. paulistorum* would not accept any courting males of both parental subspecies as well as their hybrid brothers. The hybrid females achieve this by always assuming the rejection posture displayed by nonreceptive females of *D. paulistorum.*

One final example involves the genetic control of the song produced by male crickets to attract mates (see Chapters 6 and 7). The calling song of crickets is species-specific and considered to be beneficial, in that females can recognize conspecific males and not waste time and energy responding to males of other species. The specificity of cricket songs involves a combination of fundamental frequency and the temporal pattern of the procession of pulses known as phrase structure (Figure 12–3). Bentley and Hoy (1972) analyzed the songs of two Australian field crickets, *Teleogryllus commodus* and *T. oceanicus,* and reciprocal hybrids raised in the laboratory. Figure 12–4 shows oscillograms that reveal the distinctly different characteristics of hybrid calls and the two

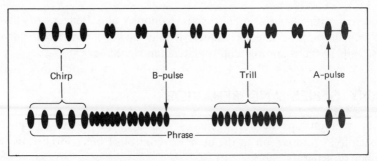

Figure 12–3. Phrase structure of the song of two species of the cricket genus *Teleogryllus*. Each phrase is composed of A-pulses that comprise the chirp portion and B-pulses that comprise the trill; *T. oceanicus* above, *T. commodus* below. (From D. R. Bentley and R. R. Hoy, 1972, *Anim. Behav.*, **20:** 478–492.)

parental calls. Analysis of a number of calls showed a prominent intertrill interval in the song of *T. oceanicus* that is absent from the song of *T. commodus*. The same intertrill interval is present in the call of hybrids from the *T. oceanicus* female to *T. commodus* male cross but absent from the call of *T. commodus* female to *T. oceanicus* male hybrids. This suggested that the presence or absence of the intertrill interval is associated with the X chromosome the males receive from their mother. It is interesting that the hybrid females preferred the songs of the hybrid males over those of either of the parental species.

Figure 12–4. Oscillograms of the songs. (A) *Teleogryllus oceanicus*, (D) *Teleogryllus commodus* and their reciprocal hybrids, (B) *T. oceanicus* female × *T. commodus* male, and (C) *T. commodus* female × *T. oceanicus* male. The beginnings of the next phrase is indicated by an arrow and 0.5 second is indicated by the horizontal time mark. (From D. R. Bentley and R. R. Hoy, 1972, *Anim. Behav.*, **20:** 478–492.)

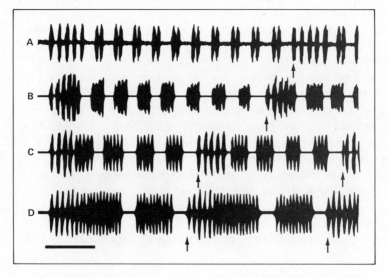

Since the intertrill interval is apparently sex-linked, and other characteristics of the call of the hybrids are intermediate between the parental calls and therefore indicate polygenic inheritance, it can be concluded that the temporal pattern of the song is controlled by an unknown number of autosomes as well as by the X chromosome (Hoy, 1974).

BEHAVIOR, EVOLUTION, AND SPECIATION

The experimental demonstration that many patterns of behavior are inherited was an important step in the development of modern evolutionary theory. Since any inherited trait is subject to natural selection, we must anticipate some evolutionary change in innate patterns of behavior. If we apply the modern interpretation of evolution, which includes a change in the frequency of occurrence of different genes in a population as a result of changes in the expected number of surviving offspring a species produces, it is relatively simple to imagine how some changes in innate behavior become established.

Arthropods in general, and insects in particular, are ideal subjects for the examination of the inheritance of behavior because their stereotyped "fixed action patterns" are so apparent, and it is often relatively easy to determine when a change occurs. There are, of course, different degrees of variation in patterns of behavior that can be acted upon by changing selective pressures, just as there are for morphological and physiological traits. Furthermore, the study of the evolution of insect behavior is not complicated by the learning of aberrant behaviors from parents and kin and the resultant indirect passage of such traits to subsequent generations.

Since selective pressures are present in the environments of all species, we can conclude that any pattern of behavior that somehow alters the impact of the environment has the potential to modify one or more selective forces and thereby influence the course of a species' evolution. A number of simple examples can be given. Insects display basic responses to stimuli that enable them to orient toward favorable zones in their habitat; a simple change in certain fixed action patterns could result in an individual ending up in an unfavorable microhabitat, where it might perish and thereby terminate the existence of the aberrant gene(s). Alternatively, an individual could end up in a more favorable area, in which more offspring and consequently more of its genes may be left. In either case there would be some change in the gene frequency associated with that particular pattern of behavior. Behavioral changes are more likely to lead to some modification of the impact of the physical environment than structural changes. Since the harshness and variability of the physical environment poses major problems to terrestrial species, it is not difficult to determine how changes in behavior that result in nest construction would be favored.

Frequently, there is a close correlation between behavior and morphology, as observed in the audio communication of the two *Gryllatalpa* species discussed earlier, and the benefits are readily identified. There are, however, both behavioral and morphological traits that do not seem to contribute to the survival of the individual that possesses them. Ritualized combat and various forms of altruistic behavior, such as the release of an alarm signal, constitute two examples that will be examined in more detail a little later.

Behavior and fitness

Any genetically based changes in behavior are acted upon by various selective pressures such that there may be either an increase or reduction in the number of progeny left by a parent, and a corresponding change in the probability with which the parental traits will be transmitted. This has been amply demonstrated by studies of mutations in *Drosophila* that alter normal courtship and mating behavior. For example, the males of a number of *Drosophila* mutants display different levels of success in their stimulation of females and the speed with which they can entice them into copulation. Various eye and body pigment mutations affect mate location and visual stimulation during courtship; these factors, in turn, affect the frequency and rate of different matings and the probability with which the mutant traits are passed on.

Provided that a trait or inseparable group of traits is not otherwise deleterious, improved mating success should increase the rate with which a genotype is transmitted. All other things being equal, we could expect the offspring of the fast-mating pairs to have a better opportunity to establish, since they would not have to compete with the offspring of the slow-mating pairs. Fast-mating males could also mate more frequently and therefore have more influence upon the genetic make-up of the next generation than slow-mating males. On the basis of these assumptions, we could expect the population to evolve toward more rapid mating.

Unfortunately, it has been more difficult to identify behavioral traits under the control of single genes than has been the case for morphological traits. We might expect that many specific responses are under the control of a single gene, but, since most fixed action patterns involve a number of integrated responses, they are usually under the control of a number of genes. However, the few examples of a behavioral trait under the control of a single gene that have been discovered have contributed greatly to our understanding of behavior and fitness. Logically, we would expect a behavioral trait that is governed by a single gene to act like a switch mechanism in determining whether or not a certain behavior occurs. However, in his study of nest-cleaning behavior mentioned earlier, Rothenbuler (1964) determined that such is not necessarily the case. He found that heterozygote, nonhygienic bees (Uu) do engage in hygienic activities (uncapping), but at a low frequency and only when the stimulus is sufficiently strong.

Although genes that act like a switch mechanism can produce rather dramatic changes in the thresholds for some kinds of response, we more frequently encounter small, more subtle, quantitative changes in behavior as a result of a point mutation. For example, the **yellow** mutant males of *Drosophila melanogaster* display essentially the same courtship behavior as normal males but are less stimulating to their mates. Apparently, the fixed action pattern of wing vibration of the mutant males is the same as that of normal males, but they do it less frequently (Bastock, 1956).

Many patterns of behavior that have a profound influence on an insect's fitness, such as host selection, are extremely complex, and their genetic control is difficult to analyze. Yet it is hypothetically possible for a single-gene mutation to cause a change in behavior that would lead to a host change, as we will see a little later on. However, some change in one or more genes that would permit survival on the new host might be necessary for the host change to persist (Bush, 1974). That is, although a change in host selection behavior may appear to improve a species' chances to survive by releasing it from competition on its original host or by increasing its resource options, it may not actually improve fitness unless there are modifications in other genes that would assure survival. In other words, there must be a positive relationship between several fitness factors for any one, like a change in behavior, to be evolutionarily significant. In the examples given we can identify some pronounced differences in the relationship of a change in behavior to fitness. On one hand, a difference in the change in male mating behavior may, on its own, be an important component of fitness because no other change is necessary to modify the gene pool in subsequent generations. Conversely, a seemingly beneficial change in host selection behavior may not be sustainable in the absence of other changes and therefore would reduce fitness.

Ritualized combat

In Chapter 7 we discussed several aspects of ritualized combat associated with the defense of mating territories and the winning of a mate. Often, these forms of behavior do not progress to an expected conclusion but, instead, lead to the establishment of a victor without serious injury to either individual. This would appear to be more of an advantage to the species or group than to the individual, since the fact that both males survive would increase the likelihood of more females becoming mated. However, from the point of view of the individual, it would seem more advantageous if its adversary were killed; the vanquished would be unable to return and fight again and would not leave offspring to compete for limited resources with the offspring of the survivor. Nevertheless, evolution has indeed moved toward combat without injury.

Maynard Smith (1978) provided a unique way to explain how ritualized combat has become evolutionarily stable, that is, how it has become resistant to the establishment of mutant strategies in the population. In a conflict, an individual either can be aggressive and escalate a contest until a solution results from serious injury to one combatant or be passive and flee whenever the opponent escalates the contest. If an arbitrary but logical payoff or penalty is assigned to the various possible outcomes of a battle, it is possible to determine the relative merits of various patterns of response. Obviously, if two individuals display aggressive behavior and continue to escalate the conflict, one ultimately will be seriously injured or killed whereas the other will be victorious. If two individuals respond passively, the result might be a conflict settled only after a long encounter at a considerable cost in time and energy to both parties. However, mixed strategy encounters would result in a rapid solution with a positive payoff for the aggressive individual and no cost or payoff (a neutral result) to the passive individual. From the assignment of numerical payoffs and the application of game theory, Maynard Smith concluded that the mixed strategy is the most evolutionary stable (Figure 12–5).

Maynard Smith admits that the preceding analysis is somewhat of an oversimplification even though conflicts of this type probably do exist. More commonly, the interactions between two individuals will involve some degree of asymmetry. For example, one individual may benefit from being the possessor of a resource; the familiarity with a territory frequently gives an occupant some advantage over an interloper. In at least

Figure 12–5. A model of ritualized combat applying game theory. The values (payoffs) assigned to various kinds of outcome are shown at the upper left. Calculations of the result of encounters between the possible combinations of behavioral types are shown at the upper right. The resulting payoffs are shown in the matrix below. According to Maynard Smith encounters between unlike behavioral types are the most evolutionarily stable. (From "The evolution of behavior" by Maynard Smith, copyright © 1978 by *Scientific American, Inc.* All rights reserved.)

Serious injury = −20
Victory = +10
Long contest = −3

$E(H, H) = \frac{1}{2}(+10) + \frac{1}{2}(-20) = -5$
$E(H, D) = +10$
$E(D, H) = 0$
$E(D, D) = \frac{1}{2}(+10) + (-3) = +2$

	Hawk (H)	Dove (D)
Hawk (H)	−5	+10
Dove (D)	0	+2

some cases, the advantage held by the possessor of a resource may be more abstract than real and gives rise to what Maynard Smith calls the bourgeois strategy. Under this strategy, an individual behaves aggressively when in possession of a resource but otherwise behaves passively (see Davies, 1978).

BEHAVIOR, REPRODUCTIVE ISOLATION, AND SPECIATION

Speciation occurs when the gene flow between diverging populations is reduced to a level at which genes entering the population as a result of hybridization are eliminated by natural selection; speciation is usually completed by the development of prezygotic isolating mechanisms. According to Bush (1975), the origin of new species of sexually reproducing organisms involves either allopatric, parapatric, or sympatric development of reproductive isolation. Allopatric or geographic speciation, which usually results from either the subdivision of a population into two large subpopulations separated by a barrier or the establishment of an isolated population in a new area (colonization), has been considered the most common mode of speciation. Parapatric speciation occurs when new species develop at the point of contact between contiguous populations; genetically, this may resemble a second form of allopatric speciation but without spatial isolation.

Neither allopatric nor parapatric speciation has a change in behavior as a major component. Such is not the case in sympatric speciation, which requires the development of reproductive isolation within a population where gene flow is not limited by either a physical barrier or reduced encounters at its periphery; some form of behavioral change would seem to be the most likely mode of reproductive isolation under these circumstances. Bush (1975) suggests that sympatric speciation appears to be limited to phytophagous and zoophagous parasites, since they exploit a genetically controlled habitat and host selection mechanisms that could be subject to change. We noted in Chapter 8 that many insects do indeed display distinct behavioral preferences for habitats and more specifically, hosts. Since roughly two thirds of all insects are parasites (herbivores being considered as parasites of plants), we can speculate that sympatric speciation, and therefore behavior, has played a major role in the development of insect diversity.

Frequently, the preferences of insects involve responses to rather specific stimuli, such as a single chemical cue. In a diverse community, the large variety of stimuli present could lead to a large number of specific relationships. In the extraordinarily diverse *Drosophila* fauna of the Hawaiian Islands, for example, we find many closely related species with seemingly slight differences in their preferences for food and resting places; Dobzhansky et al. (1956) noted the differential attraction of *Drosophila* species to different species of yeast. Ehrman and Parsons (1976) review some of the rather subtle behavioral differences between similar species of *Drosophila* that can act as isolating mechanisms; in the case of the sibling species *D. pseudoobscura* and *D. persimilis,* they list the following factors that keep them isolated in regions where they are sympatric.

1. The two species have somewhat different habitat preferences; *D. persimilis* prefer cooler areas than *D. pseudoobscura.*
2. The two species have different food preferences, including differential attraction to different yeasts.
3. The two species were differently attracted to baits in the morning and evening, including differences in peak activity periods.
4. *D. persimilis* had a greater affinity for light than did *D. pseudoobscura.*
5. The males of the two species have different courtship songs.

The first four factors listed would not necessarily result in complete reproductive isolation, but they do illustrate how even a slight change in behavior could tend to divide a population and at least open the way to speciation.

We do not therefore need to think in terms of a genetic revolution as a prerequisite to sympatric speciation. Even though the behavioral aspects of habitat and host selection may involve complex interactions between an insect and a large array of environmental cues under the control of many gene loci, it is possible for a single-gene change to produce a significant change in habitat or host selection behavior. A single-gene mutation could change the affinity of a specialized receptor and thereby alter the interpretation of, and the response to, a stimulus. Since many parasites use their hosts as a rendevous site for courtship and mating, a slight change in host selection behavior could provide the first step toward reproductive isolation. Such is the case for fruit fly *Rhagoletis.*

The native American hawthorn fly, *Rhagoletis pomonella,* has developed new host races on introduced European fruits, such as apples and cherries. As a result of a long-term study of *Rhagoletis,* Bush (1969) developed a tentative model for sympatric speciation that clearly illustrates the importance of behavior as an isolating mechanism. Figure 12–6 summarizes the host selection and mating behavior of these insects. Bush (1969) concludes that for species of the *pomonella* group, the most important reproductive isolating mechanism is host plant preference.

Once a behavioral change has occurred that makes possible a host change, other genetic modifications may be necessary before the new host

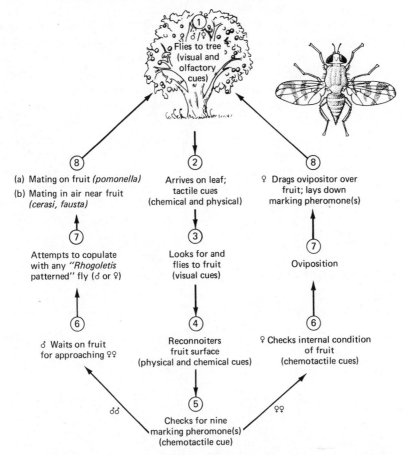

Figure 12–6. Host selection and mating behavior of the fruit fly *Rhago-letis*. (Redrawn from G. L. Bush, 1974, in M. J. D. White, ed., *Genetic Mechanisms of Speciation in Insects,* Australia and New Zealand Book Co., Sydney.)

relationship can be fully established. Bush (1974) identifies three more or less discrete genetic components involved in the complete adaptation to a new host; these are genes affecting host recognition, survival, and inducement to oviposit on the new host. Nonetheless, the process begins with a change in behavior.

If a parasite becomes established on a new host, it may exert a strong selective pressure that could, in turn, lead to the evolution of a new defensive mechanism by the host. The insect may then evolve a new counterdefense, some aspect of which could be behavioral. This ongoing coevolution usually tends to increase the reproductive isolation between the old and new races and thereby hasten speciation.

Another method by which a change in behavior could lead to sympatric speciation would be through a change in the chemical structure of a reproductive pheromone, on one hand, and the response of some individuals to it, on the other. Carde et al. (1975) discovered two populations of the European corn borer, *Ostrinia nubilalis,* that are different in the phero- mone used in communication and do not interbreed where they occur sympatrically. The males of the eastern populations are attracted to the more common female sex pheromone consisting of 96 percent *trans*–11– tetradecenyl and 4 percent *cis*–11–tetradecenyl, whereas males of the western population respond to the same isomers, but with the proportions reversed. The difference in responsiveness and the resulting reproductive isolation could have arisen as a result of two independent mutations, one altering the males *trans* receptor to a *cis* receptor, and another altering the females enzyme system to produce the *cis* rather than the *trans* isomer (Bush, 1975).

CHAPTER 13

Eusocial behavior

Whereas parental care and various kinds of presocial behavior are common throughout most of the insect orders, eusocial behavior is restricted to the Isoptera (termites) and Hymenoptera (ants, and some wasps and bees). Eusocial insects are characterized by the possession of three characteristics: cooperation among individuals in the care of the young, a division of labor in which more or less sterile individuals assist fecund individuals, and an overlap of generations so that offspring are able to assist their parents (Wilson, 1971; Michener, 1974). The termites hold the honor of antiquity among eusocial species, but I will discuss the eusocial Hymenoptera first because the wasps display the behavioral sequence used in Chapter 10 to provide some insight into the evolution of social behavior.

EUSOCIAL WASPS

All of the eusocial wasps belong to the family Vespidae. Among these, the genera *Polistes* (paper wasps), *Vespa* (hornets), and *Vespula* (yellow jackets) are the best known. Each of these genera have three castes: **queens** (fertile females), **workers** (nonfertile females), and **drones** (males). In temperate regions, wasp colonies are established each spring by over-wintered queens and last but a single season. When a queen comes out of hibernation, she immediately begins the construction of a nest and rears the first small group of daughters. In the genus *Polistes,* a founding queen may be joined by other overwintered queens, but these auxiliary queens are always subordinate to the foundress, which eats any eggs her helpers might lay (Eberhard, 1969). Normally, however, wasp colonies are started by a single foundress that constructs an initial tier of hexagonal paper cells, rears the first group of larvae on macerated insects, and keeps the small nest clean. The first group of offspring are all workers, which upon emergence add new cells to the nest, tend the queen, and forage for food. From this time on, the founding queen rarely leaves the nest and spends the rest of her life laying eggs.

The colony grows rapidly throughout the summer as each generation of workers increases the size of the nest and the number of young that can be reared at one time (Figure 13–1). Toward the end of summer, some larger brood cells are constructed in which a number of males and new queens are reared. When the young queens and drones emerge, they leave the nest to mate and never return. The founding queen dies, and the colony declines to nothing as the remaining workers expire.

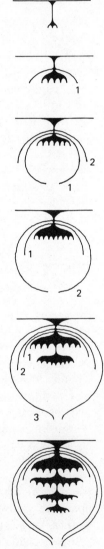

Figure 13–1. The steps in the construction of a nest by a vespid wasp. (Adapted from H. Kemper and E. Döhring, 1967, *Die sozialen Faltenwespen Mitteleuropas*, Paul Parey, Berlin.)

EUSOCIAL BEES

The eusocial bees consist of a number of species with a variety of social life styles. Among them the well-known bumblebees and honeybees represent quite different levels of social complexity, whereas the stingless bees of the genera *Trigona* and *Malipona* display an interesting combination of social complexity and primitive nesting behavior.

The bumblebees are primarily adapted to cooler climates. Like the eusocial wasps, they normally overwinter as queens that survive the cold months in the solitude of protected hibernation sites. In the spring, the queen emerges and sets out in search of a suitable subterranean nest site in the form of some preexisting cavity. Within the nest she constructs a cuplike cell from wax that is secreted by the intersegmental glands of her abdomen. The cup is then provisioned with a ball of pollen, on which she lays her first batch of eggs, and the cell is sealed. After the initial egg cell has been completed, the queen constructs a honeypot near the entrance to the nest and fills it with nectar. The first group of offspring are all workers that upon emergence assist the queen in producing more larval cells and in feeding the new brood. The subsequent brood is either fed by the workers or is allowed to feed directly on a mass pollen store, depending on the species. By late summer, the colony consists of several hundred workers and produces males and new queens; the colony then abandons its nest. The virgin queens are met by waiting males, copulate, and then enter hibernation.

The tropical, stingless bees construct their wax and resin nests in hollow tree trunks or roots or in subterranean crevices. The nest is usually divided into a brood area and food storage area. The brood combs consist of horizontal tiers of cells that open upward and are constructed one above the other like an upside-down wasp nest. Each tier is begun with a central cell to which rings of peripheral cells are added until each comb reaches its full diameter. Unlike the wasps, stingless bees construct one cell at a time, provide it with an egg, then provision and seal it before the next cell is built. Unlike the honeybees, which use their combs over and over again, the stingless bees tear down their cells after the young emerge. The food storage area consists of a series of wax pots that are filled with honey and pollen.

Stingless bee colonies expand to thousands of individuals and, unlike those of bumblebees, persist from year to year. As with the honeybees, stingless bees produce maiden queens and drones, from time to time, and the number of colonies is increased and dispersed by swarming. Unlike honeybee swarms, however, stingless bee swarms consist of a cluster of workers and a maiden queen rather than the old queen.

The social life of the honeybee, *Apis mellifera,* is far more complex, as was indicated by the discussion of its rather elaborate communication system in Chapter 6. The honeybee originated in the tropics or subtropics but seems to have been preadapted, through its ability to regulate its nest temperature, to expand its range into regions of colder climate and still maintain perennial colonies. Entire colonies, instead of just fecund queens as in the wasps and bumble bees, overwinter. This not only permits the colonies to become much larger but permits and requires other changes as well.

Because honeybee queens do not initiate new colonies each year by their own labor, they do not require any of the capabilities of their worker caste. Consequently, they have become much more differentiated from the workers morphologically and serve solely the functions of reproduction. New colonies are founded through the process of colony division by swarming, as in the stingless bees. The swarming process usually begins when a colony is strong and crowded. In response to some unknown stimulus, the queen reduces the level of queen substance she produces, and the workers respond by constructing a small number of queen cells. The larvae that hatch from the eggs laid in these larger cells are fed exclusively on royal jelly and complete their development in about two thirds of the time it takes workers and drones to develop. When virgin queens are developing within the colony the resident queen's egg production declines, and she is treated less hospitably until more or less forced to leave with a large retinue of older workers. Swarms containing the old queen are called prime swarms and may be followed a short time later by one or more afterswarms, each containing a new queen produced within the hive and already mated.

Swarms usually settle in a temporary cluster while a permanent nesting site is located. Once scout bees have found and communicated the location of a suitable nesting place, the temporary cluster breaks up and moves to the new site. The workers immediately begin to construct vertical sheets of hexagonal wax cells suspended from the roof of the nest site. The wax combs (Figure 13–2) are constructed a precise distance apart and consist mainly of regular, well-fitted worker cells. The size and shape of the cells determine the kind of egg the queen lays therein and how each cell is provisioned. Worker cells receive diploid eggs and are provisioned with royal jelly for a few days, followed by a mixture of pollen and honey called bee bread. When a queen encounters an irregular, enlarged cell, she will lay a haploid egg, which becomes a drone. When an egg is laid in a cuplike cell, the workers will feed the larvae that develops only on royal jelly and thereby produce maiden queens. Thus the structural details of the wax comb play an integral part in the social organization of the colony.

Since the queen that accompanies a primary swarm is mated and the workers can immediately begin constructing a brood comb, the queen can

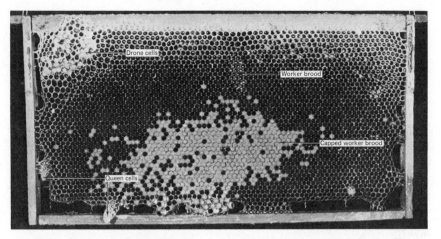

Figure 13–2. A frame of honey comb from a commercial hive showing worker cells, drone cells, queen cells, and some capped brood. (Courtesy of University of California.)

soon proceed with egg laying, and the new colony grows very quickly. Meanwhile, the virgin queen left with the original colony must go on one or more nuptial flights to obtain enough sperm to last the 5 to 7 years that she might live. The original colony also grows rapidly, since the brood left by the departed queen emerges and gradually takes over the hive chores from the aging workers that remained after the swarm.

Drones are produced throughout the more favorable part of the year. They are usually most abundant when the colonies are reproducing by swarming and therefore contain virgin queens that must be inseminated. Drones contribute absolutely no labor to the colony but are tolerated by the workers until late in the summer. As the supply of food in the field begins to decline in the fall, the workers become intolerant of the drones, driving them from the hive and not permitting them to reenter.

In respect to brood rearing and overwintering, the ability to regulate nest temperature is highly beneficial. The precision with which honeybee colonies do this is truly remarkable, but it would not be possible if it were not for the type of nest site selected and the way it is modified and utilized. The actual thermoregulatory process, however, is accomplished by the bees themselves rather than physically, as with the termites. During the months when a brood is being reared, the brood chamber is maintained at a temperature of 34.5 to 35.5° C, even though the outside air temperature may reach more than 60° C. In the winter, the cluster temperature ranges from 20 to 30° C and is never permitted to drop below 17° C, regardless of the outside temperature. When the outside temperature drops, the temperature in the hive is maintained by the metabolic heat generated by the workers, whose behavior changes as the temperature declines. At first, the workers form a loose cluster toward the center of the hive. As the temperature drops, the cluster tightens. The workers at the center consume small amounts of honey and generate heat by vibrating their muscles; those forming the outer layers of the cluster act as an insulating blanket. As time passes, the outer bees move toward the center of the cluster, while those near the center move into the outer layers. During hot weather, the temperature of the brood area is maintained by a circulation of air in the hive created by workers that fan with their wings at the hive entrance. When this activity proves to be inadequate, other workers carry water into the hive and distribute it over the brood cells, which are then cooled as the water evaporates.

ANTS

All ants are eusocial. The lack of presocial behavior of the kinds found among the wasps and bees can be explained by the fact that the ants probably evolved from a group of presocial wasps. The variety of life styles displayed by the ants has long been of fascination to naturalists, perhaps because so many aspects of their social behavior seemingly parallel our own. The brief account that follows will scarcely scratch the surface of this intriguing subject, so the reader whose interest is aroused should examine either E. O. Wilson's *The Insect Societies* or W. M. Wheeler's *Ants, Their Structure Development and Behavior*.

All living species of ants have a caste system that generally parallels that of the eusocial wasps and bees and clearly reveals a division of labor. The principal castes consist of males, queens, and workers, the latter group being subdivided into different types in most species. Several genera also have castes intermediate between males and workers and between workers and queens, but these are generally less important than the main castes. Typical males are winged reproductives that contribute absolutely nothing to the labor of the colony although they do groom other adults and are groomed in return. Queens are the colony founders and mothers. They are usually winged when they emerge but shed their wings after their nuptial flight. During the establishment of a new nest, the queen performs all of the tasks later performed by the workers; but, once the first group of workers has emerged, the queen's activities are reduced to egg laying and grooming.

Workers are sterile females, and in the broad sense include both the laborers and defenders of the colony. There are often three subcastes of workers, based partially on size (Figure 13–3). The largest are the **majors** or **soldiers**, which are often considered to be a distinct caste because of the oversized (allometric) development of their heads and mandibles. The other workers are distinguished as being **media**, if of inter-

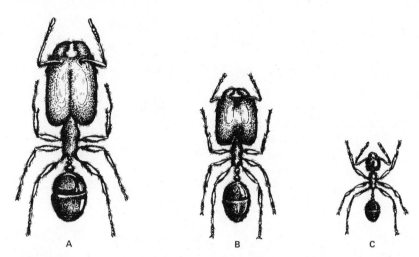

Figure 13–3. Different forms of the worker cast of ants. (A) Major. (B) Media. (C) Minor. (After W. M. Wheeler, 1910, *Ants: Their Structure Development and Behavior,* Columbia University Press, New York.)

mediate size, and **minors**, if small. The degree of worker polymorphism varies greatly from species to species, but, in species in which it is pronounced, a clear division of labor among the subcastes can be observed. The most unusual workers are the repletes of the honey ant *Myrmecocystus.* These individuals serve as casks for the storage of the honeydew brought to the nest by the foragers. When filled, the abdomen of a replete becomes a large swollen sac.

Ants are extremely numerous and ubiquitous. Their food habits are diverse, although most are omnivorous, and their social groups range in size from hundreds to perhaps millions. Certainly, the most spectacular of these groups are the legionary or army ants of the humid tropical forests. These ants do not construct nests but form temporary clusters called bivouacs in the shelter of a fallen tree trunk or other such partially exposed location. The workers form a solid mass up to a meter across, consisting of layer upon layer of individuals linked together by their tarsal claws. The queen, thousands of larvae and pupae, and, at certain times of the year, up to 1000 males and a few virgin queens, are located near the center of the ball of up to a half million workers.

The night is passed in a tight cluster, but, at first light, the chains of workers become detached and a hoard of individuals begins to move away from the bivouac site in all directions. Soon, one or more raiding columns form and move out in search of food. The workers lay a pheromone trail to guide those that follow, while the soldiers guard the column's flanks. Some species engage in column raids, whereas others employ a swarming tactic (Figure 13–4). Regardless of their method, however, the foraging ants form a number of changing columns behind the advancing hoard of

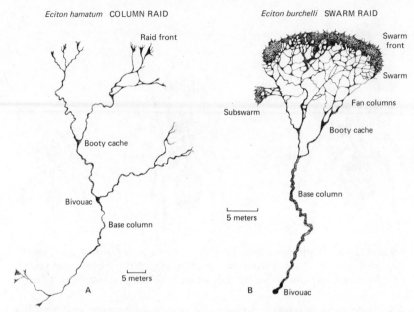

Eciton hamatum COLUMN RAID   *Eciton burchelli* SWARM RAID

Figure 13–4. Basic raiding patterns employed by army ants. (A) The column raid of *Eciton hamatum* with an advancing front composed of several narrow columns of workers. (B) The swarming raid of *Eciton burchelli* composed of a mass of workers advancing together. (From E. O. Wilson, 1971, after Rettenmeyer, 1963, *The Insect Societies,* The Belknap Press of Harvard University Press, Cambridge, Mass., with permission of the author and Harvard University Press.)

workers that flushes and captures a variety of mainly orthropod prey. The prey are stung to death, dismembered, and transported rearward as food for the developing larvae (Rettenmeyer, 1963). At the end of each day of marauding, the protective cluster is reformed.

As the colony rapidly reduces the food supply in the immediate vicinity of its bivouac, it must move to a new location. However, in at least one species, the change in bivouac site has been correlated with the reproductive cycle within the colony, which in turn has an effect on its food requirements. According to Schnierla (1971), the queen periodically undergoes rapid ovarian development, which results in great distension of her abdomen. Over a period of several days, such a queen lays 100,000 to 300,00 eggs, which hatch about 3 weeks later. During this **statary phase**, the larvae from the previous period of oviposition complete their pupal development and emerge as young workers. The addition of these new workers seems to stimulate the colony to increase the intensity of its daily foraging raids and change the location of its bivouac site each evening. This **nomadic phase** therefore coincides with the period during which a large number of larvae must be fed. Once these larvae complete their development and pupate, the nomadic behavior declines, and the colony enters another reproductive period, during which a single bivouac site is utilized.

People living in the temperate zone are most familiar with the nest-building ants, such as members of the genus *Myrmica,* which usually nest in the ground under a large stone, or the genus *Formica,* which construct the familiar mounds of twigs and other plant debris. These ants are characterized by the production of winged reproductives during the summer and early fall. In some species the emergence of the reproductives produces quite spectacular swarms, during which mating occurs. The nuptial flight results in the ubiquitous spread of potential foundresses of new colonies. After a short time, the newly mated queen sheds her wings and excavates the beginning of a new nest into which she seals herself. In the spring, she lays a small group of eggs and tends the larvae until they attain adulthood and become the colony's initial group of workers. These first offspring may take more than a year to complete their development, but the queen never leaves, relying on her internal food reserves and flight muscles for her own sustenance as well as that of her young.

Over the succeeding years, the colony grows very slowly until enough workers are present for it to enter a period of accelerated growth. In areas characterized by cold winters, the colonies enter hibernation for several months, and no eggs are laid until warmer conditions prevail in the spring. In most species, several years pass before winged males and females are produced to leave and establish new colonies in their turn. Colonies of many species contain only a few hundred individuals, whereas some contain many thousands; most nests contain only one queen, but some *Formica* nests have hundreds of active queens.

In spite of the general similarities displayed by the nest-building ants, as far as the establishment and growth of new colonies is concerned, they show extensive adaptive radiation in other aspects of their behavior. Most display a high level of variability in their choice of prey. Like the army ants, they forage widely for almost any kind of small animals, especially arthropods, that are available. A single nest of the European red ant, *Formica polyctena,* is said to gather up to a kilogram of such food in the course of a single day. Chauvin (1967) concluded that the estimated population of 300 million million red ants, which inhabit the Italian Alps, would be capable of destroying 15,000 tons of insects a year. But not all ants have the catholic taste of *F. polyctena.* Some feed specifically on certain kinds of arthropods or their eggs. Others, often called "harvester" or "agricultural" ants, subsist entirely on seeds, and thereby become pests in some grain- and grass-growing regions. When protein-rich foods are not needed for the rearing of larvae, many ants feed upon the nectar of flowers.

A number of ant species feed exclusively on the anal excretion of aphids, scale insects, and other homopterans. The so-called "honeydew" is rich in sugars and free amino acids and forms an inexhaustible source of nutrient throughout the time when the producing insects are actively feeding. Some species of ants can be seen to stroke the cornicles of aphids to induce the flow of honeydew from the anus. Ants, in their tending of homopterans, can seriously interfere with the biological control programs directed against these insects by actively repelling parasites and predators. In the fall, some ants actually transport aphids or their eggs from their host plant to underground "barns" in which they are cared for over the winter. In the spring, the ants carry the aphids back to their host plant once again to "milk" them of their honeydew (Way, 1963).

Members of the New World tribe Attini are sophisticated fungus culturers. Many species in the group gather small pieces of green leaves or the petals of flowers to form underground beds on which they culture a specific variety of fungus. Members of the genus *Atta* are among the more serious pests of New World tropical agriculture as a result of their leaf collecting; in English, they are commonly referred to as leaf-cutting ants. Media workers carry the plant material piece by piece deep into their nests, where it is licked, cut into small pieces, wet with an anal secretion, and formed into beds of moist pulp. The newly formed beds are then "planted" with fungal mycelia collected from established beds. The fungus grows rapidly and soon produces on the tips of the hypae small spheres, which are fed to the ant larvae. The fungus beds are tended and harvested by smaller workers (minora) that never engage in the collection of leaf fragments. However, in the case of *Atta cephalotes,* these smaller workers may accompany the larger leaf-gatherers; they do not assist with the leaf cutting but ride back to the nest on the leaf portion, apparently warding off parasitic phorid flies with their mandibles and hind legs (Figure 13–5).

One of the more fascinating aspects of the fungus culture is how the ants are able to maintain a monoculture of their specific fungus; abandoned beds are rapidly overgrown by alien fungi of various species. Weber (1957) discovered that the worker ants tending the fungus beds "weed" them of alien hypae with their mandibles. It has also been postulated, but not proved, that the ants employ fungicidal and bacteriocidal substances secreted by their salivary or anal glands. The cultured fungus is carried to each new nest by the foundress. Before departing on her nuptial flight, the virgin queen packs a small wad of mycelia into a cavity near the base of her labium. The wad is then deposited in the nest she excavates. The queen tends the initial fungus garden but does not consume any of the culture herself. However, the first workers to emerge feed upon it and fertilize it with their fecal material.

Not all ants are as industrious as the aphid herders and fungus gardeners. The Amazon slaver ant, *Polyergus lucidus,* and the red ant, *Formica sanguinea,* are common species that make raids on other ant species to capture slaves. The raiders carry larvae and pupae back to their nests,

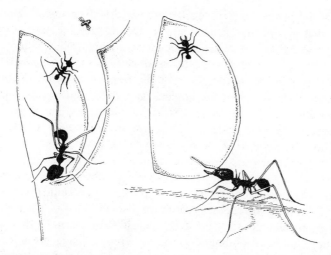

Figure 13–5. Media worker of the fungus-growing ant, *Atta cephalotes,* cutting and carrying a portion of leaf while being protected against attack from a phorid fly by a minor worker. (Redrawn from I. Eibl-Eibesfeldt and E. Eibl-Eibesfeldt, 1967, *Z.für Tierpsych.,* **24:** 278–281.)

where they are raised to assume the duties of nest maintenance. *Polyergus lucidus* is an obligatory slave keeper, as the workers cannot feed themselves. *F. sanguinea,* on the other hand, is a facultative slaver (Wilson, 1975).

TERMITES

The order Isoptera is clearly more ancient and more primitive than the order Hymenoptera, yet all living termites are eusocial and have achieved a level of caste differentiation and social organization equal to the wasps, ants, and bees. The most primitive living termite, *Mastotermes darwiniensis,* seems closely related to the primitive, social, wood-eating roach *Crytocercus,* with which they share a number of intestinal protozoans that are essential to the digestion of the cellulose they consume in a wide variety of forms. Some termites also lay eggs in small groups that are protected by a covering similar to the oothecae of roaches.

Termites lack cellulose-digesting enzymes and rely on their intestinal fauna of mutualistic flagellate protozoans, which must be passed continuously between individuals, to aid in the breakdown of their food. The passage of gut symbionts from old to young individuals seems to have set the stage for the development of more advanced social behavior. It has been postulated, therefore, that the social order of termites began with trophallaxis and led to the addition of brood care, rather than the reverse, as in the social hymenoptera.

The natural history of termites is quite variable. The so-called dry wood termites that frequently cause damage to the dry, seasoned wood of buildings usually live in colonies consisting of only a few hundred individuals. Winged reproductives are produced seasonally, usually during the warmer months, leave the nest soon after emergence, and spread rather ubiquitously. When a female completes her flight, she divests herself of her wings and runs about in an excited fashion until joined by a male. After pairing, the king and queen engage in a behavior called tandem running, in which the king closely follows his queen in search of a nest site. When the pair locate a suitable place to begin a nest, they take turns excavating an initial tunnel with a small chamber at the bottom. The king and queen seal themselves in by plugging the tunnel with chewed wood, and the queen proceeds to produce a small batch of eggs. When the young nymphs emerge, they are fed regurgitated food until they are able to feed upon the surrounding wood and thereby enlarge the nest. As the colony grows in the ensuing years, soldiers are produced that provide the colony with defense. When the colony is several years old, more alates are produced, which found new colonies. The foundress of a colony may live for more than 10 years, but if her egg production declines or she is killed, her duties may be assumed by secondary queens.

Most of the more highly evolved termites are soil dwellers that construct conspicuous mound nests. Alate reproductives are again produced seasonally and often emerge from their nests at the beginning of the rainy season. After a short, feeble flight, the reproductives shed their wings and engage in nest founding behavior similar to that already described. The first brood consists of workers, but soldiers are produced later. In addition to workers, soldiers, and huge physogastric females, some colonies contain secondary and tertiary queens. As the number of individuals in the colony grows from year to year, the size of the mound is increased by the addition of layers of soil and excrement to form the most elaborate insect-built structures.

The mound-building termites have adapted to virtually every form of food with a high cellulose content. The food is gathered by the workers, which forage throughout a large area surrounding the nest by way of a system of subterranean tunnels or covered pathways; some even forage on the surface over well-trodden trails marked with pheromones. Termites of the subfamily Macrotermitinae culture fungus on comblike structures built from their excrement. As in the ant genus *Atta,* the fungus beds of occupied nests are a pure culture. Instead of cropping the fungus in the manner of ants, termites consume the entire mass, including the excrement substratum. The colonies of fungus-growing termites may contain as many as 2 million individuals. The primary queen is capable of laying up to 30,000 eggs a day; in a life span of up to 10 years, a single queen may produce tens of millions of offspring.

The gigantic mounds of tropical termites often have distinctive forms that are characteristic of particular species (Figure 13–6). In addition

Figure 13–6. Mound nests of some Australian termites. (A and B) North-south and east-west aspects of mounds of *Anutermes meridionalis*, (C) Nest of *Nasutitermes triodiae*. (D) Nest of *Nasutitermes walkeri*. (E) Nest of *Amitermes vitiosus*. (From *Insects of Australia*, 1970, with permission of the Melbourne University Press.)

to providing protection, termite nests are often constructed in such a way as to provide a well-regulated interior microclimate. *Amitermes* orient their nests in such a way as to maximize solar warming in the morning and evening (Figure 13–6A and B). Microclimatic control is most highly developed in the fungus-growing genus *Macrotermes*. Its members construct a series of passageways, through which air flows according to its own density, so that temperature and carbon dioxide conditions favorable to the growth of fungus are maintained in the culture chamber. The basic structure of the nest of *Macrotermes natalensis* of Africa and a diagram of its air conditioning system are shown in Figure 13–7.

The evolution of the termites is far removed phylogenetically from

Figure 13–7. Section through the mound nest of the African termite *Macrotermes natalensis* showing the basic design of the air conditioning tunnels. (Adapted from M. Lüscher, 1961, *Sci. Am.*, **205** (1): 138–145.)

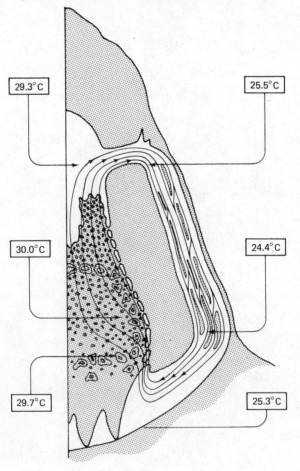

Table 13-1. Basic Similarities and Differences in Social Biology Between Termites and Higher Social Hymenoptera (Wasps, Ants, Bees); Similarities Are the Result of Evolutionary Convergence

Similarities	Differences	
	Termites	**Eusocial Hymenoptera**
1. The castes are similar in number and kind, especially between termites and ants	1. Caste determination in the lower termites is based primarily on pheromones; in the higher termites it involves sex, but some of the other factors remain unidentified	1. Caste determination is based primarily on nutrition, although pheromones play a role in some cases
2. Trophallaxis (exchange of liquid food) occurs and is an important mechanism in social regulation	2. The worker castes consist of both females and males	2. The worker castes consist of females only
3. Chemical trails are used in recruitment as in the ants, and the behavior of trail laying and following is closely similar	3. Larvae and nymphs contribute to colony labor, at least in later instars	3. The immature stages (larvae and pupae) are helpless and almost never contribute to colony labor
4. Inhibitory caste pheromones exist, similar in action to those found in honeybees and ants	4. There are no dominance hierarchies among individuals in the same colonies	4. Dominance hierarchies are commonplace but not universal
5. Grooming between individuals occurs frequently and functions at least partially in the transmission of pheromones	5. Social parasitism between species is almost wholly absent	5. Social parasitism between species is common and widespread
6. Nest odor and territoriality are of general occurrence	6. Exchange of liquid anal food occurs universally in the lower termites, and trophic eggs are unknown	6. Anal trophallaxis is rare, but trophic eggs are exchanged in many species of bees and ants
7. Nest structure is of comparable complexity and, in a few members of Termitidae (e.g., *Apicotermes, Macrotermes*), of considerably greater complexity; regulation of temperature and humidity within the nest operates at about the same level of precision	7. The primary reproductive male (the "king") stays with the queen after the nuptial flight, helps her construct the first nest, and fertilizes her intermittently as the colony develops; fertilization does not occur during the nuptial flight	7. The male fertilizes the queen during the nuptial flight and dies soon afterward without helping the queen in nest construction
8. Cannibalism is widespread in both groups (but not universal, at least not in the Hymenoptera)		

Source: E. O. Wilson, *The Insect Societies,* The Belknap Press of Harvard University Press, Cambridge, Mass., 1971.

that of the Hymenoptera; yet, the two groups have, by quite different routes, arrived at a complex level of social organization. There are some striking similarities in the social biology these distinct orders display, but there are some fundamental differences as well. The most important differences markedly affect the make-up of the colonies. In the case of the termites, the queen cannot control the sex of her offspring, so the workers are both males and females (genetically). Furthermore, after the first few instars, the termite immatures engage in work for the good of the colony. In the social Hymenoptera, the workers are all female, and the queen can regulate the production of males; the larvae are, in a sense, parasitic and contribute nothing to the well-being of the colony. These and other differences are summarized in Table 13–1 with the kind permission of E. O. Wilson.

CHAPTER 14

Practical aspects of insect behavior

In the introduction I remarked that there are pragmatic as well as intellectual reasons for studying insect behavior. Much basic behavioral information has clearly played an important role in the development of population theory and modern concepts of population management, but, in addition, specific elements of behavioral information have been sought in conjunction with the development of new approaches to the direct suppression of pest populations. In recent years, the manipulation of insect behavior has become a popular aspect of applied entomology.

As we have seen in several earlier chapters, insect behavior is for the most part stereotyped and is therefore predictable within the limitation of the normal range of variability inherent in populations (see Chapter 11). This predictability permits safer application than would be the case if the patterns of behavior were constantly changing as a result of experience. As a result, behavior plays a role in the success of almost every pest control tactic presently in use.

To attempt to present a review of the role of behavior in all of the insect control tactics and strategies now employed in agriculture, forestry, and public health would be at the very least foolhardy. Nevertheless, there are a number of situations in which behavior not only plays a key role but in which a failure to apply what is known about a species' behavior can seriously reduce the efficacy of the program. The application of a knowledge of behavior is also important to the planning and conduct of the sampling programs on which management decisions are made.

There seems to be no best way to organize a discussion of this kind because it is impossible to avoid overlap and cross referencing. Since I am not attempting a complete coverage and the primary orientation of this book is behavior, I will attempt to show the applied importance of some of the major aspects of behavior discussed earlier.

BEHAVIORAL PERIODICITY

The behavior of insects is greatly influenced by cyclical phenomena such as daily patterns of temperature change and the diurnal cycle of light and darkness. Although such changes are variable in terms of how well they can be predicted, an understanding of their influence on behavior is extremely important. Although factors of the physical environment may be somewhat variable, they may serve as the timing mechanism (*Zeitgeber*) of highly predictable circadian rhythms. The exogenous and endogenous rhythms of insects form an important aspect of management decision making. For example, we can use such knowledge to time insecticide applications to coincide with periods of pest activity or quiescence. A simplistic example involves the overnight application of pesticides during the nocturnal feeding period of caterpillars while avoiding contact with foraging honeybees.

Insects also display circadian rhythms in their sensitivity to pesticides, which, when more completely understood, could have considerable practical significance. At least in some cases, maximum sensitivity to poisons coincides with periods of peak activity and metabolism. The house crickets, *Acheta domestica,* for example, are most susceptible to several compounds during the first few hours of their active period (Nawosielski, et al., 1964). In other cases, sensitivity seems to be related to photoperiod (see Fernandez and Randolph, 1966). These data suggest that biochemical oscillations linked to overt behavioral rhythms could be exploited in pest control. Such knowledge could be used specifically to increase the compatibility of pesticide use with biological control so as to increase the effectiveness against the target while reducing the impact on beneficial species. Since chemical and biological control are two major tactics incorporated into integrated pest management programs, improving their compatability is important (Bartlett, 1964). The timing of pesticide applications to coincide with the maximum sensitivity and chance of contact with the target, while perhaps avoiding activity and susceptibility periods of parasites and predators, is a distinct possibility. Behavioral periodicity could increase the effectiveness of other control tactics as well. The endogenous control of pupal eclosion displayed by some flies could be a determining factor in the time of release of sterile males.

Many examples of the applicability of periodicity data in the management of both beneficial and injurious species could be cited, but probably there is no single aspect of applied entomology where such knowledge is more important than in sampling. In spite of the fact that we consistently observe alternating periods of activity and quiescence, as well as cyclical, fundamental changes in the behavior of many insects, there seem to be few sampling plans that take these aspects of insect biology into account. Sampling is a major tool in the evaluation of pest populations, and the results obtained from sampling usually determine whether or not a control program is implemented. Sampling methods vary greatly, from the use of traps, through some standardized use of a sweep net, to the counting of the number of individuals feeding on a unit area of host plant foliage. Often, the size of a crop unit is many acres, and more attention is paid to coverage and the elimination of edge effects than the time span of the total sampling program. Often, too, the sampling crew will work in a block throughout the day, and this raises questions concerning how meaningful the data collected are. If the insect being sampled has a diurnal activity rhythm or changes behavior throughout the day, do samples collected early in the morning equate with those collected at midday or later in the afternoon? During the day, changes in activity levels might well affect the efficiency of the sampling method, and changes in response could result in a change of position relative to the sampling universe. Influx and exodus that may occur at specific times during the day would also affect the data collected.

PATTERNS OF ORIENTATION AND DISPLACEMENT

Within the cycles of behavior governed by internal clocks and by daily changes in the physical environment, insects often display behavioral changes that result from the interplay of internal drives and the resultant changes in the response thresholds for various kinds of stimuli. A reciprocal pattern of responses of this kind was discussed in Chapter 4 as being an important element of migratory behavior. A strong migratory drive usually reduces the responsiveness to appetitive stimuli such as food and mates. In some species, the life cycle seems to be divided between these rather different activities, whereas in others they seem to be more closely intertwined. The lady beetle, *Hippodamia convergens,* completes its reproduction and feeding in the valley, migrates to the mountains, hibernates, and then migrates back to the valley to reproduce the following year. On the other hand, some bark beetle populations may be composed of both nonmigrants and migrants. Data obtained from rotating nets designed to capture flying bark beetles may show two peak periods of flight activity—one in the morning and one in the afternoon—whereas data obtained from field olfactometers containing host odors and pheromones may reveal only the afternoon peak. This would suggest that the morning flight activity is of a migratory nature, involving individuals that are not responsive to appetitive stimuli, whereas the afternoon flight is related to host location and attack.

Data of the kind described again draw attention to the importance of a knowledge of behavior in population sampling. Although some workers have suggested the use of traps baited with attractants as a means of pest population suppression, the primary value of such traps is in sampling programs. However, the fact that different portions of a population may be responsive to such traps, depending on prior events, means the usefulness of trap catch data is placed in jeopardy unless the behavior of the species being monitored is well known. To return to the bark beetle example mentioned previously, pheromone trap catches only provide an indication of changes in the population of responsive individuals. Since the behavioral make-up of the population (migrants versus responders) changes throughout the season and from year to year, it is not possible to accurately determine the size of the total population from the number of beetles trapped.

In addition to the cyclic activity patterns associated with internal clocks and regular changes in the physical environment, insects capable of activity may display changes in behavior or their level of activity in relation to changes in the intensity of cues used in orientation. Male moths, for example, may rest until stimulated into activity by the reception of a female's sex attractant. In Chapter 5 I discussed the importance of polarized light to insect behavior as shown by the investigations of Wellington (1974a; b). If Wellington's conclusions concerning flight and territorial behavior at midday can be applied fairly generally to diurnal species, then polarized light should be considered in a variety of insect management programs. Certainly, if most insects are unable to traverse open spaces when polarized light is absent from the overhead sky, midday would be a poor time to sample aerial populations over the uniform landscape that characterizes many agroecosystems. On the other hand, the period of relative inactivity around solar noon may be the best time to make parasite and predator releases, as it would promote some local searching for hosts or prey before dispersing, perhaps to areas where food species are less concentrated (Wellington, 1974a). Wellington (1976) used his polarized light studies to illustrate the general need to apply behavioral studies to solving entomological problems.

COMMUNICATION AND REPRODUCTIVE BEHAVIOR

Obviously, the knowledge we have gained about insect communication has suggested several potentially useful pest control tactics. The use of insect chemical messengers stands out as being one of the most active areas of pest control research. Unfortunately, the technical chemical aspects of the field have advanced more rapidly than our understanding of some of the relevant behavioral aspects. As pointed out earlier, attractants can be extremely useful in sampling programs, provided they take into account behavioral periodicity and variability and they are not used to estimate population density. Much more efficient detection programs for quarantined species, such as the Japanese beetle and several fruit flies, have been made possible through the use of attractants.

The use of attractants to reduce pest populations by trapping out large numbers of individuals has been proposed, and pilot projects have been undertaken (Figure 14–1). However, the efficacy of such programs is difficult to evaluate. The number of individuals captured may be unrelated to the size of the local population prior to trapping if insects are attracted over a long distance or if an influx of individuals continues to replace those trapped out. In at least one test, however, significant population reduction was achieved; a population of codling moth, as well as fruit damage, was reduced substantially using pheromone traps in an apple orchard (MacLellan, 1976). Perhaps the greatest benefits will accrue in the use of attractants by combining them with poisons, chemosterilants, or pathogens. The Oriental fruit fly, *Dacus dorsalis,* was eradicated from the island of Rota through the distribution of small cards impregnated with an attractant and an insecticide. A potential problem with the use of programs involving attractants is the possible decline of parasite and predator populations. Not only does the trapped portion of the population represent a reduction in the food resources of natural enemies, but also the phenomenon of host and prey location by beneficials, using the formers' chemical communication, could result in a direct reduction in the numbers of the pests' natural enemies.

The antithesis of attraction is, of course, repulsion. The problem with this as a population or damage reduction technique is that it does not directly reduce the population and may, in fact, increase it elsewhere. This does suggest, however, that it may be possible to suppress a population with repellents or chemicals that produce confusion, thereby increasing in-flight mortality and time of exposure to natural enemies.

The success of the sterile male technique used to control the screw-worm fly and several fruit flies was dependent upon an understanding of basic reproductive behavior. One of the important factors in population reduction using this method lies in the capacity of sterilized males to compete in all ways with normal wild males. Whenever a sterile male program is undertaken, it is vital to examine the behavior of treated males carefully. In addition to being vigorous, they must not display changes in courtship or mating behavior that would cause females to reject them.

The use of auditory communication in insect management has lagged behind the use of chemical communication but has some potential. Haskell (1964) indicated that sound could be used to destroy unwanted insects, but the general use of ultrasonics would probably not be feasible in that it would be injurious to other forms of life. However, insect pro-duced sound could be exploited for pest detection, and it may be possible to use simulated insect sounds as either attractants or repellents.

Adams and associates (1953) used the sound produced by feeding

Figure 14–1. A large number of bark beetles captured on a sticky trap baited with a vial containing a mixture of attractant compounds. This type of trap has been used in pilot projects designed to reduce local beetle populations. (Courtesy of Gary Pitman, Oregon State University.)

weevils to detect infestations in grain, and it may be possible to apply a similar technique to the detection of structural pests. The use of sound as a control tactic has some potential but will probably remain limited in scope. Belton (1962) conducted field experiments with sonic repellents and, in one situation, obtained a 50 percent reduction in an infestation of the European corn borer, *Ostrinia nubilalis*. However, as with all control tactics involving repellency, the population may move elsewhere and any resultant mortality would be difficult to evaluate.

One practical application of sound in the management of a beneficial species involves monitoring hive sounds as a means of predicting honeybee swarms. A monitoring device called an *apidictor* was developed by Woods (1959) to reduce the time necessary to open and examine hives for swarm prevention.

HOST SELECTION BEHAVIOR

Host selection is probably the single most important aspect of insect behavior from an applied point of view. In spite of the controversies and knowledge gaps that seem to characterize this area, host selection by insects lies at the very heart of agricultural and medical entomology.

As indicated in Chapter 8, the chemical and physical characteristics of plants and animals are a major factor in the ease with which hosts are located, identified, and chosen—they form the basis of preference or nonpreference. Crop plant selections must be made both in terms of their desirability as human food and in terms of pest preference. Often, the natural compounds that are components of a plant's defensive arsenal produce undesirable flavor characteristics as far as the consumer is concerned. Yet, when these factors are selected against in plant breeding, they may increase the desirability of the variety to insects. On the other hand, characteristics that deter insect attack without affecting marketability are sought by plant breeders developing pest-resistant strains.

In the biological control of both insects and weeds, there is a great deal of dependence placed upon the reliability of stereotyped host selection behavior that results in an identifiable degree of host specificity. Obviously, the breadth of the host range must be determined, and there must be a high level of confidence that there will not be any transfer to beneficial or nontarget species. If, for example, a herbivore imported to control a weed was not specific and subsequently transferred to a desirable plant species, a new pest problem could ensue.

Host specificity is an important characteristic of entomophagous insects released as biological control agents. Those that have a narrow host range usually display population characteristics that quite closely parallel those of their host, whereas less-specific agents may have a tendency to switch from a less-abundant food source to a more-abundant one. This may result in the population of the target species not being suppressed far enough to inhibit a rapid population rebound. On the other hand, highly specific biological control agents may be eliminated from a local area by a lack of hosts, making it necessary to reintroduce the agent or await natural repopulation.

The established methodology of biological control usually involves the importation, behavioral screening, mass rearing, and subsequent release of the parasite or predator, followed by a period of field evaluation for indications of establishment and success. Behavior screening in the laboratory may not reveal the true nature of the host selection preferences that can play a major role in their success as control agents. In the laboratory, parasites and predators do not have to search for their hosts and prey in the same manner as in the field. The first phase of host-finding behavior often involves the location of the host's habitat by identification of characteristics, such as the odor of the host's host plant. When the pest is a polyphagous herbivore like the cabbage looper, *Trichoplusia ni,* or the gypsy moth, *Lymantria dispar,* the level of biological control achieved may only be adequate when the pest occupies the habitat containing those plants that tend to be most attractive to the control agent.

One cannot help but wonder how many potentially valuable biological control agents have been more or less discarded because they were evaluated in a portion of an ecosystem in which they did not display their full host-finding capability. Furthermore, the mass rearing programs used to generate parasites and predators often involve the use of the most easily obtained plant material or an artificial diet for the host rearing aspect. This procedure could lead to some degree of behavioral conditioning or a change in host selection behavior that might reduce the effectiveness of parasites following their release in the wild.

When considering the selection of host plants by herbivores, we tend to think mainly in terms of the attractant characteristics even though the selection process may involve an interplay of attraction and repulsion. In general, the repulsion of insects is a short-range phenomenon that may do little more than inhibit landing and the initiation of feeding. For this reason, repellents have been used mainly to protect people and domestic animals against species that cause irritation, or vector pathogens. However, chemicals extracted from plants have proved to be strong feeding deterrents under laboratory conditions. Smissen and associates (1957) and Maxwell and associates (1965) have discouraged the feeding of the European corn borer and the Colorado potato beetle. but more work is needed to develop a method for use under field conditions.

PARENTAL CARE

The principal form of brood care behavior that influences pest control involves the placement of the eggs and the manner in which they are

Figure 14–2. Mobile field domicile for the leaf-cutter bee *Megachile rotundata.* The shelter contains frames filled with boxes of drinking straws as shown in the lower photograph. (Courtesy of William Stephen, Oregon State University.)

protected. Eggs laid in the open are readily accessible to ovicide sprays, whereas those deposited in the soil, in plant tissue, or subsequently covered by some protective coating may be more difficult to destroy. The timing of oviposition, as well as location, often influences the modification of cultural practices for pest control purposes. The planning of tillage, irrigation, and the removal of plant debris can all be beneficially planned with a knowledge of oviposition behavior.

An understanding of nest construction and brood care has been a major contributing factor in the development of practices for the management of the alkali bee, *Nomia melanderi,* and the leaf-cutter bee, *Megachile rotundata,* for pollination. Studies of nest site preferences and the capability of brood-rearing females to tolerate dense aggregations of nests were important to the design of mobile domiciles (Figure 14–2) and permanent bee beds (see Stephen, 1960; 1962).

Probably, a knowledge of behavior has contributed more to the management of the honeybee, *Apis mellifera* than of any other single insect. Not long ago the honeybee was managed in a rather haphazard manner, primarily for the production of honey and to a lesser extent wax. Studies on the brood care behavior of the honeybee led to the design of the modern removable frame hive and new management practices that simplify swarm control, permit commercial pollen, royal jelly, and queen production, and improve crop pollination.

CONCLUSION

The foregoing represent only a few of the many examples that reveal the importance of a knowledge of behavior to the development and modification of programs for the management of insects. Much is yet to be learned that should aid significantly in the formulation of new management strategies as well as in the improvement of strategies already in use.

Although we have described insect behavior as being largely stereotyped, we should not equate this with simplicity. Insect behavior is extremely complex. One may be tempted to jump to the application of a new or interesting response discovered in the laboratory, but this can be risky in that insect behavior tends to be a complex interplay of many responses. It is necessary to learn how each response fits into the sequence of responses that comprises the over-all pattern of behavior. These patterns are adaptations for survival that have evolved over long periods of time, and we must expect them to contain certain checks and balances that act as buffers against some degree of external change. Frequently, we desire to apply our knowledge of insect behavior to situations quite different from those in which the pattern of behavior evolved. If we fail to take into account the adaptive significance of behavior under the conditions in which it evolved, we may find that we cannot apply what we have learned as effectively as we would wish. Let us examine the migratory behavior of the lady beetle, *Hippodamia convergens,* as an example.

When the new generation of young adult beetles leaves the crop lands of California's San Joaquin Valley, the vegetation is still green, and, unless insecticides have been applied, populations of aphids on which the lady beetles feed are still flourishing. However, the valley is now extensively irrigated and bears little resemblance to the ecosystem in which the migratory behavior of *H. convergens* evolved. Prior to the introduction of intensive agriculture to the valley, the native vegetation dried up by early summer and the aphid populations declined. On this basis we should expect the seemingly premature departure of the lady beetles. Similarly, we should not have great expectations for the biological control potential of beetles collected from hibernating aggregations in the mountains unless they are in some way preconditioned to enter the reproductive phase of their life cycle.

Behavioral variability is the other important consideration that must be kept in mind when we attempt to apply behavior to management. No matter how we exploit insect behavior, we cannot assume it will be effective in terms of every member of a population. If some portion of the variation is consistently unaffected, we can expect it to increase and perhaps dominate the population if the selective pressure we impose is continuous. For example, the use of pheromone traps would select out responsive individuals, in a way that could eventually lead to a fundamental change in population behavior. We must therefore pay attention to the relative importance to a population of the individuals removed and those left behind. As Wellington (1977) indicated, we should become more concerned about qualitative changes in populations as a result of a management practice than quantitative changes. The quality of an individual can be measured in many different ways, but behavioral traits can provide the key to identifying them.

Cited references

Abbot, C. H. 1951. A quantitative study of the migration of the painted lady butterfly, *Vanessa cardui* L. *Ecology,* **32**:155–171.

Adams, R. W., J. E. Wolfe, M. Milner, and J. A. Shellenberger. 1953. Aural detection of grain infested internally with insects. *Science,* **118**:163–164.

Adkisson, P. L., and S. H. Roach. 1971. A mechanism for seasonal discrimination in the photoperiodic induction of pupal diapause in the bollworm, *Heliothis zea* (Boddie). Pages 272–280 *in* M. Menaker, ed., *Biochronometry,* National Academy of Sciences, Washington, D.C.

Alexander, R. D. 1961. Aggressiveness, territoriality, and sexual behavior in field crickets (Orthoptera: Gryllidae). *Behaviour,* **17**:130–223.

Alexander, R. D. 1975. Natural selection and specialized chorusing behavior in insects. Pages 35–78 *in* David Pimentel, ed., *Insects, Science and Society.* Academic Press, Inc., New York.

Alloway, T. M. 1972. Learning and memory in insects. *Ann. Rev. Entomol.,* **17**:43–56.

Atkins, M. D. 1961. A study of the flight of the Douglas-fir beetle *Dendroctonus pseudotsugae* Hopk. (Coleoptera: Scolytidae) III. Flight capacity. *Can. Entomol.,* **93**:467–474.

Atkins, M. D. 1966. Laboratory studies on the behavior of the Douglas-fir beetle, *Dendroctonus pseudotsugae* Hopk. *Can. Entomol.,* **98**:953–991.

Atkins, M. D. 1967. The effect of rearing temperature on the size and fat content of the Douglas-fir beetle. *Can. Entomol.,* **99**:181–187.

Atkins, M. D. 1969. Lipid loss with flight in the Douglas-fir beetle. *Can. Entomol.,* **101**:164–165.

Bartlett, B. R. 1964. Integration of chemical and biological control. Pages 480–511 *in* P. DeBach, ed., *Biological Control of Insect Pests and Weeds.* Van Nostrand Reinhold Company, New York.

Bastock, M. 1956. A gene mutation which changes a behavior pattern. *Evolution,* **10**:421–439.

Beard, R. L. 1963. Insect toxins and venoms. *Ann. Rev. Entomol.,* **8**:1–18.

Beling, I. 1929. Über das Zeitgedachtnis der Bienen. *Z. vergl. Physiol.,* **9**:259–338.

Belton, P. 1962. Responses to sound in pyralid moths. *Nature,* **196**:1188–1189.

Bennet-Clark, H. C. 1970a. The mechanism and efficiency of sound production in mole crickets. *J. exp. Biol.,* **52**:619–652.

Bennet-Clark, H. C. 1970b. A new French mole cricket, differing in song and morphology from *Gryllotalpa gryllotolpa* L. (Orthoptera: Gryllotalpidae). *Proc. R. Entomol. Soc. London* (*B*), **39**:125–132.

Bentley, D. R., and R. R. Hoy. 1972. Genetic control of the neural network generating cricket (*Teleogryllus gryllus*) song patterns. *Anim. Behav.*, **20**:478–492.

Benzer, S. 1973. Genetic dissection of behavior. *Sci. Am.*, **229**:24–37.

Blake, G. M. 1958. Diapause and the regulation of development in *Anthrenus verbasci* (L.) (Coleoptera, Dermestidae). *Bull. Entomol. Res.*, **49**:751–775.

Blake, G. M. 1959. Control of diapause by an "internal clock" in *Anthrenus verbasci* (L) (Coleoptera, Dermestidae). *Nature*, **183**:126–127.

Blest, A. D. 1957. The function of eyespot patterns in the Lepidoptera. *Behaviour*, **11**:209–256.

Bösiger, E. 1967. La signification évolutive de la sélection sexuelle chez les animaux. *Scientia*, **102**:207–223.

Bradbury, W. C., and G. F. Bennett. 1974. Behavior of adult Simuliidae (Diptera). I. Response to color and shape. *Can. J. Zool.*, **52**:251–259.

Brower, L. P., J. V. Z. Brower, and E. P. Cranston. 1965. Courtship behavior of the queen butterfly *Danaus gilippus*. *Zoologica*, **50**:1–39.

Brown, R. G. B. 1966. Courtship in the *Drosophila obscura* group. II. Comparative studies. *Behaviour*, **25**:281–323.

Brown, W. A. 1966. The attraction of mosquitoes to hosts. *J. Am. Med. Assoc.*, **1966**:249–252.

Brun, R. 1914. Die Raumorientation der Ameisen und das Orientierungsproblem in allgemeinen. Gustav Fischer, Jena, Austria.

Buck, J. B. 1948. The anatomy and physiology of the light organ in fireflies. In *Bioluminescence Ann. N.Y. Acad. Sci.*, **49**:397–482.

Bush, G. L. 1969. Sympatric host race formation and speciation in frugivorous flies of the genus *Rhagoletis* (Diptera, Tephritidae). *Evolution*, **23**:237–251.

Bush, G. L. 1974. The mechanism of sympatric host race formation in the true fruit flies (Tephritidae). Pages 3–23 *in* M. J. D. White, ed., *Genetic Mechanisms of Speciation in Insects*. Australia and New Zealand Book Co., Sydney.

Bush, G. L. 1975. Modes of animal speciation. *Ann. Rev. Ecol. Sys.*, **6**:339–364.

Butler, C. G. 1960. The significance of queen substance in swarming and supersedure in honey bee (*Apis mellifera* L.) colonies. *Proc. Roy. Entomol. Soc. London (A)*, **35**:129–132.

Butler, C. G. 1967. Insect pheromones. *Biol. Rev.*, **42**:42–87.

Cade, W. 1975. Acoustically orienting parasitoids: Fly phonotaxis to cricket song. *Science*, **190**:1312–1313.

Campbell, R. W. 1979. Gypsy moth: Forest influence. U.S. Department of Agriculture, Forest Service. Agriculture Information Bulletin No. 423.

Carde, R. T., J. Kochansky, J. F. Stimmel, A. G. Wheeler, Jr., and W. L. Roelofs. 1975. Sex pheromones of the European corn borer *Ostrinia nubilalis: cis* and *trans* responding males in Pennsylvania. *Environ. Entomol.*, **4**:413–414.

Chapman, J. A. 1954. Studies on summit frequenting insects in western Montana. *Ecology*, **35**:41–49.

Chapman, J. A. 1967. Response behavior of scolytid beetles and odour meteorology. *Can. Entomol.,* **99**:1132–1136.

Chapman, R. F. 1969. *The Insects: Structure and Function.* American Elsevier Publishing Co., Inc., New York.

Chapman, R. F. 1974. The chemical inhibition of feeding by phytophagous insects: A review. *Bull. Entomol. Res.,* **64**:339–363.

Chapman, R. F., and J. S. Kennedy. 1976. General conclusions. Pages 307–309 in T. Jermy, ed., *The Host-Plant in Relation to Insect Behavior and Reproduction.* Plenum Publishing Corporation, New York.

Chauvin, R. 1967. *The World of an Insect.* World University Library, London; McGraw-Hill Book Company, New York.

Collett, T. S., and M. F. Land. 1975. Visual spatial memory in a hoverfly. *J. Comp. Physiol.,* **100**:59–84.

Cook, W. C. 1967. Life history, host plants, and migrations of the beet leaf-hopper in the western United States. *Tech. Bull. U.S. Dept. Agr.,* No. 1365.

Corbet, P. S. 1966. The role of rhythms in insect behavior. Pages 13–28 in P. T. Haskell, ed., *Insect Behavior. Roy. Entomol. Soc., London Symposium 3.*

Corbet, P. S., C. Longfield, and N. W. Moore. 1960. *Dragonflies.* Collins Clear-Type Press, London.

Crystal, M. M. 1964. Observations on the role of light, temperature, age, and sex in the response of screw-worms to attractants. *J. Econ. Entomol.,* **57**:324–325.

Davies, N. B. 1978. Territorial defence in the speckled wood butterfly (*Pararge aegeria*): The resident always wins. *Anim. Behav.,* **26**:138–147.

DeBach, P. 1964. *Biological Control of Insect Pests and Weeds.* Van Nostrand Reinhold Company, New York.

Debaisieux, P. 1938. Organes scolopibiaux des pattes d'insectes. *Cellule,* **47**:77–202.

DeRuiter, L. 1952. Some experiments on the camouflage of stick caterpillars. *Behaviour,* **4**:222–232.

Dethier, V. G. 1947. *Chemical Insect Attractants and Repellants.* McGraw-Hill Book Company, New York.

Dethier, V. G. 1954. The physiology of olfaction in insects. *Ann. N.Y. Acad. Sci.,* **58**:139–157.

Dethier, V. G. 1963. *The Physiology of Insect Senses.* John Wiley & Sons, Inc., New York.

Dethier, V. G. 1966. Feeding behavior. Pages 46–58 in P. T. Haskell, ed., *Insect Behavior. Roy. Entomol. Soc., London Symposium 3.*

Dethier, V. G. 1970. Chemical interactions between plants and insects. Pages 83–102 in E. Sondheimer and J. B. Simione, eds., *Chemical Ecology.* Academic Press, Inc., New York.

Dethier, V. G., L. B. Brown, and C. N. Smith. 1960. The designation of chemicals in terms of the responses they elicit from insects. *J. Econ. Entomol.,* **53**:134–136.

Dixon, A. F. G. 1973. *Biology of Aphids,* Edward Arnold & Co., London.

Dobzhansky, T., D. M. Cooper, H. J. Phaff, E. P. Knapp, and H. L. Carson.

1956. Studies on the ecology of *Drosophila* in the Yosemite region of California. 4. Differential attraction of species of *Drosophila* to different species of yeasts. *Ecology,* **37**:544–550.

Downes, J. A. 1958. Assembly and mating in the biting Nematocera. *Proc. 10th Int. Congr. Entomol.,* Montreal (1956) **2**:425–434.

Downes, J. A. 1969. The swarming and mating flight of Diptera. *Ann. Rev. Entomol.,* **14**:271–298.

Eberhard, Mary Jane West. 1969. The social biology of polistine wasps. *Miscellaneous Publications, Museum of Zoology,* University of Michigan, Ann Arbor, **140**:1–101.

Ehrlich, P. R., and P. H. Raven. 1964. Butterflies and plants: A study in coevolution. *Evolution,* **18**:586–608.

Ehrlich, P. R., and P. H. Raven. 1967. Butterflies and plants. Pages 195–202 *in* T. Eisner and E. O. Wilson, eds., *The Insects. Scientific American,* 1977. W. H. Freeman and Company, Publishers, San Francisco.

Ehrman, L. 1960. A genetic constitution frustrating the sexual drive in *Drosophila paulistorum. Science,* **131**:1381–1382.

Ehrman, L. 1961. The genetics of sexual isolation in *Drosophila paulistorum. Genetics,* **46**:1025–1038.

Ehrman, L., and P. A. Parsons. 1976. *The Genetics of Behavior.* Sinauer Associates Inc., Sunderland, Mass.

Eibl-Eibesfeldt, I., and E. Eibl-Eibesfeldt. 1967. Das Parasitenabwehren der Minima-Arbeiterinnen der Blattschneider-Ameise (*Atta cephalotes*). *Z. Tierpsychol.,* **24**:278–281.

Eisner, T. 1970. Chemical defense against predation in arthropods. Pages 157–217 *in* E. Sondheimer and J. B. Simeone, eds., *Chemical Ecology.* Academic Press, Inc., New York.

Eisner, T., and Y. C. Meinwald. 1966. Defensive secretions of arthropods. *Science,* **153**:1341–1350.

Evans, H. E. 1958. The evolution of social life in wasps. *Proc. 10th Int. Congr. Entomol.,* Montreal, 1956. **2**:449–457.

Evans, H. E. 1973. *Wasp Farm.* Anchor Natural History Books, Anchor Press, Doubleday & Company, Inc., New York.

Evans, W. G. 1975. Wax secretion in the infrared sensory pit of *Melanophila acuminata* (Coleoptera: Buprestidae). *Quaest. Entomol.,* **11**:587–589.

Evans, W. G .1976. Circadian and circatidal locomotory rhythms in the intertidal beetle, *Thalassotrechus barbarae* (Horn): Carabidae. *J. Exp. Mar. Biol. Ecol.,* **22**:79–90.

Fernandez, A. T., and N. M. Randolph. 1966. The susceptibility of houseflies reared under various photoperiods to insecticide residues. *J. Econ. Entomol.,* **59**:37–39.

Fitzegrald, T. D. 1973. Coexistence of three species of bark-mining *Marmara* (Lepidoptera: Gracillariidae) on green ash and descriptions of new species. *Ann. Entomol. Soc. Am.,* **66**:457–464.

Fraenkel, G. S. 1953. The nutritional value of green plants for insects. *Proc. 9th Int. Congr. Entomol.,* Amsterdam, 1951, **2**:90–100.

Fraenkel, G. S. 1959. The chemistry of host specificity of phytophagous insects. *Fourth Int. Congr. Biochem. Biochemistry of Insects.* Pergamon Press, Inc., Elmsford, N.Y.

Fraenkel, G. S. 1959. The raison d'être of secondary plant substances. *Science,* **129**:1466–1470.

Fraenkel, G. S., and D. L. Gunn. 1961. *The Orientation of Animals.* Dover Publications, Inc., New York.

Friend, W. G., and J. J. B. Smith. 1977. Factors affecting feeding by blood-sucking insects. *Ann. Rev. Entomol.,* **22**:309–331.

von Frisch, K. 1950. Die Sonne als Kompass im Leben der Bienen. *Experientia.,* **6**:210–221.

von Frisch, K. 1962. Dialects in the language of the bees. Pages 241–246 *in* T. Eisner and E. O. Wilson, eds., *The Insects. Scientific American,* 1977. W. H. Freeman and Company, Publishers, San Francisco.

von Frisch, K. 1967. *The Dance Language and Orientation of Bees.* Trans. by L. E. Chadwick. The Belknap Press of Harvard University Press, Cambridge, Mass.

von Frisch, K., and M. Lindauer. 1954. Himmel und Erde in Konkurrenz bei der Orientierung der Bienen. *Naturwissenschaften,* **41**:245–253.

Frost, S. W. 1959. *Insect Life and Insect Natural History,* 2nd ed. Dover Publications, Inc., New York.

Geier, P. W. 1960. Physiological age of codling moth females *Cydia pomonella* (L.) caught in bait and light traps. *Nature,* **185**:709.

Ghent, A. W. 1954. An investigation of the feeding behavior of the jack pine sawfly, *Neodiprion banksianae* Roh. Master's thesis, University of Toronto, Toronto, Ontario.

Glick, P. A. 1939. The distribution of insects, spiders and mites in the air. *Tech. Bull. U.S. Dept. Agr.,* No. 673.

Goldsmith, T. H., and G. D. Bernard. 1973. The visual system of insects. Pages 165–272 *in* Morris Rockstein, ed., *The Physiology of Insecta.* Academic Press, Inc., New York.

Goodman, J. L. 1970. The structure and function of the insect dorsal ocellus. *Adv. Insect Physiol.,* **7**:97–195.

Gould, J. L. 1975. Honeybee recruitment: The dance language controversy. *Science,* **189**:685–693.

Gould, J. L., J. L. Kirschvink, and K. S. Deffeyes. 1978. Bees have magnetic remanence. *Science,* **201**:1026–1028.

Gross, A. O. 1913. The reactions of arthropods to monochromatic lights of equal intensities. *J. Exp. Zool.,* **14**:457–514.

Hagen, B. W., and M. D. Atkins. 1975. Between generation variation in the fat content and behavior of *Ips paraconfusus* Lanier. *Z. ang. Entomol.,* **79**:169–172.

Hagen, K. S. 1962. Biology and ecology of predaceous Coccinellidae. *Ann. Rev. Entomol.,* **7**:289–326.

Hamilton, W. D. 1963. The evolution of altruistic behavior. *Am. Natur.,* **97**:354–356.

Hamilton, W. D. 1964. The genetical evolution of social behavior. *J. Theor. Biol.,* **7**:1–52.

Harbach, R. E., and J. R. Larsen. 1977. Humidity behavior and the identification of hygroreceptors in the adult mealworm, *Tenebrio molitor.* *J. Insect Physiol.,* **23**:1121–1134.

Hardy, A. C., and P. S. Milne. 1938. Studies in the distribution of insects by aerial currents. Experiments in aerial tow-netting from kites. *J. Anim. Ecol.,* **7**:199–229.

Haskell, P. T. 1964. Sound production. Pages 563–608 *in* M. Rockstein, ed., *The Physiology of Insecta,* Vol. 1. Academic Press, Inc., New York.

Haskell, P. T. 1966. Flight behavior. Pages 29–45 *in* P. T. Haskell, ed., *Insect Behaviour. Roy. Entomol. Soc. London Symposium 3.*

Haskell, P. T., and J. E. Moorhouse. 1963. A blood-borne factor influencing the activity of the central nervous system of the desert locust. *Nature, London,* **197**:56–58.

Haynes, D. L. 1973. Population management of the cereal leaf beetle. Pages 232–240 *in* P. W. Geier, L. R. Clark, D. J. Snderson, and H. A. Nix, eds., *Insects: Studies in Population Management. Ecol. Soc. Aust. (Memoirs 1),* Canberra.

Hocking, B. 1971. Blood-sucking behavior of terrestrial arthropods. *Ann. Rev. Entomol.,* **16**:1–26.

Hodgson, E. C. 1973. Chemoreception. Pages 127–165 *in* M. Rockstein, ed., *The Physiology of Insecta.* Vol. 1. Academic Press, New York.

Hölldobler, B. 1971a. Homing in the harvester ant *Pogonomyrmex badius. Science,* **171**:1149–1151.

Hölldobler, B. 1971b. Communication between ants and their guests. Pages 262–270 *in* T. Eisner and E. O. Wilson, eds., *The Insects. Scientific American* 1977. W. H. Freeman and Company, Publishers, San Francisco.

Hölldobler, B., and E. O. Wilson. 1977. Colony-specific territorial pheromone in the African weaver ant *Oecophylla longinoda* (Latreille). *Proc. Nat. Acad. Sci. USA,* **74**:2072–2075.

Horridge, G. A. 1962. Learning of leg position by the ventral nerve cord in headless insects. *Proc. Roy. Soc. London (B),* **157**:33–52.

Horridge, G. A., and C. Giddings. 1971. *Proc. Roy. Soc. London B. Biol. Sci.* **179**:87.

Howse, P. E. 1964. The significance of the sound produced by the termite *Zootermopsis angusticollis* (Hagen). *Anim. Behav.,* **12**:284–300.

Hoy, R. R. 1974. Genetic control of acoustic behavior in crickets. *Am. Zool.,* **14**:1067–1080.

Hummel, H., and P. Karlson. 1968. Hexansaure als Bestandteil des Spurpheromons der Termite *Zootermopsis nevadensis* Hagen Hoppe-Seyler's *Z. physiol. Chem.,* **349**:725–727.

Hutchinson, G. E. 1957. Concluding remarks. *Cold Spring Harbor Symp. Quant. Biol.,* **22**:415–427.

Jacobson, M. 1965. *Insect Sex Attractants.* John Wiley & Sons, Inc., New York.

Jander, R. 1957. Die optische Richtungsorientierung der Roten Waldameise (*Formica rufa* L.). *Z. vergleich. Physiol.,* **40**:162–238.

Jander, R. 1963. Insect orientation. *Ann. Rev. Entomol.,* **8**:94–114.

Jeannel, R. 1960. *Introduction to Entomology.* Trans. by Harold Oldroyd. Hutchinson Publishing Group, London.

Jermy, T. 1966. Feeding inhibitors and food preference in chewing phytophagous insects. *Entomol. Exp. Appl.,* **9**:1–12.

Jermy, T., ed. 1976. *The Host Plant in Relation to Insect Behaviour and Reproduction.* Plenum Publishing Corporation, New York.

Johnson, C. G. 1960. A basis for a general system of insect migration and dispersal by flight. *Nature,* **186**:348–350.

Johnson, C. G. 1963a. Physiological factors in insect migration by flight. *Nature,* **198**:423–427.

Johnson, C. G. 1963b. The origin of flight in insects. *Proc. Roy. Entomol. Soc. London (C),* **28**:26–27.

Johnson, C. G. 1969. *Migration and Dispersal of Insects by Flight.* Methuen & Co., Ltd., London.

Johnson, D. L. 1964. The adult behavior of *Libellula saturata* (Uhler) (Odonata:Libellulidae). Master's thesis, San Diego State University, San Diego, Calif.

Jones, J. C. 1968. The sexual life of a mosquito. Pages 71–78 *in* T. Eisner and E. O. Wilson, eds., *The Insects. Scientific American,* 1977. W. H. Freeman and Company, Publishers, San Francisco.

Kahn, A. A., H. I. Maibach, and W. G. Strauss. 1968. The role of convection currents in mosquito attraction to human skin. *Mosquito News,* **28**:462–464.

Karlson, P., and A. Butenandt. 1959. Pheromones (ectohormones) in insects. *Ann. Rev. Entomol.,* **4**:39–58.

Kemper, H., and E. Döhring. 1967. *Die sozialen Faltenwespen Mitteleuropas.* Paul Parey, Berlin.

Kennedy, J. S. 1975. Insect dispersal. Pages 103–119 *in* David Pimentel, ed., *Insects, Science and Society.* Academic Press, Inc., New York.

Kennedy, J. S., and C. O. Booth. 1951. Host alteration in *Aphis fabae* Scop. I. Feeding preferences and fecundity in relation to the age or kind of leaves. *Ann. Appl. Biol.,* **38**:25–64.

Kettlewell, H. B. D. 1961. The phenomenon of industrial melanism in Lepidoptera. *Ann. Rev. Entomol.,* **6**:245–262.

Kieckhefer, R. W., D. A. Dickmann, and E. L. Miller. 1976. Color responses of cereal aphids. *Ann. Entomol. Soc. Am.,* **69**:721–724.

Knerer, C., and C. E. Atwood. 1973. Diprionid sawflies: Polymorphism and speciation. *Science,* **179**:1090–1099.

Konopka, R. J., and S. Benzer. 1971. Clock mutants of *Drosophila melanogaster. Proc. Nat. Acad. Sci. USA,* **68**:2112–2116.

Laarman, J. J. 1955. The host-seeking behaviour of the malaria mosquito *Anopheles maculipennis atroparvus. Acta Leidensia,* **25**:1–144.

Lindauer, M. 1961. *Communication Among Social Bees.* Harvard University Press, Cambridge, Mass.

Lindauer, M. 1967. Recent advances in bee communication and orientation. *Ann. Rev. Entomol.,* **12**:439–470.

Lindauer, M. 1971. *Communication Among Social Bees.* Harvard University Press, Cambridge, Mass.

Lindauer, M., and H. Martin. 1968. Die Schwereorientierung der Bienen unter dem Einfluss des Erdmagnetfeldes. *Z. vergl. Physiol.,* **60**:219–243.

Lindsley, E. G., J. W. MacSwain, and P. H. Raven. 1963. Comparative behavior of bees and Onagraceae. II. *Oenothera* bees of the Great Basin. *Univ. Calif. Publ. Entomol.,* **33**:25–58.

Lloyd, J. L. 1971. Bioluminescent communication in insects. *Ann. Rev. Entomol.*, **16**:97–122.

Loftus, R. 1976. Temperature-dependent dry receptor on antennae of *Periplaneta:* Tonic response. *J. Comp. Physiol.*, **111**:153–170

Lundgren, Lennart. 1975. Natural plant chemicals acting as oviposition deterrents on cabbage butterflies (*Pieris brassicae* (L.), *P. rapae* (L.), and *P. napi* (L.), *Zoologica Scripta*, **4**:253–258.

Lüscher, M. 1961. Air-conditioned termite nests. *Sci. Am.*, **205**(1):138–145.

McElroy, W. D. 1964. Insect bioluminescence. Pages 463–508 *in* M. Rockstein, ed., *Physiology of Insecta*. Academic Press, Inc., New York.

Macfadyen, A. 1957. *Animal Ecology: Aims and Methods,* 2nd ed. Sir Isaac Pitman & Sons Ltd., London.

McIver, S. B. 1975. Cuticular mechanoreceptors of arthropods. *Ann. Rev. Entomol.*, **20**:381–397.

MacLellan, C. R. 1976. Suppression of codling moth (Lepidoptera:Olethreutidae) by sex pherome trapping of males. *Can. Entomol.*, **108**:1037–1040.

McMullen, L. H., and M. D. Atkins. 1962. On the flight and host selection of the Douglas-fir beetle *Dendroctonus pseudotsugae* Hopk. (Coleoptera: Scolytidae). *Can. Entomol.*, **94**:1309–1325.

Maldonado, H., and J. C. Barros-Pita. 1970. A fovea in the praying mantis. I. Estimation of the catching distance. *Z. vergl. Physiol.*, **67**:58–78.

Manning, A. 1956. Some aspects of the foraging behavior of bumble bees. *Behaviour*, **9**:164–201.

Manning, A. 1966. Sexual behavior. Pages 59–68 *in* P. T. Haskell, ed., *Insect Behaviour. Roy. Entomol. Soc. London Symposium 3*.

Markl, H. 1962. Schweresinnes-organe bei Ameisen und anderen Hymenopteren. *Z. vergl. Physiol.*, **44**:475–569.

Markl, H., and M. Lindauer. 1965. Physiology of insect behavior. Pages 3–122, *in* M. Rockstein, ed., *The Physiology of Insecta*, Vol. 2. Academic Press, Inc., New York.

Maxwell, F. G., W. L. Parrott, J. N. Jenkins, and H. N. Lafever. 1965. A boll weevil feeding deterrent from the calyx of an alternate host, *Hibiscus syriacus. J. Econ. Entomol.*, **58**:985–988.

Maynard Smith, J. 1978. The evolution of behavior. *Sci. Am.*, **239**(3):176–192.

Michener, C. D. 1974. *The Social Behavior of Bees: A Comparative Study*. The Belknap Press of Harvard University Press, Cambridge, Mass.

Milne, L. J., and Margery Milne. 1976. The social behavior of burying beetles. *Sci. Am.*, **235**(2):84–89.

Nault, L. R., L. J. Edwards, and W. E. Styer. 1973. Aphid alarm pheromones: Secretion and reception. *Environ. Entomol.*, **2**:101–105.

Nawosielski, J. W., R. L. Patton, and J. A. Naegele. 1964. Daily rhythm of narcotic sensitivity in house cricket, *Gryllus domesticus* L., and the two-spotted spider mite, *Tetranychus urticae* Koch. *J. Cell. Comp. Physiol.*, 63:393–398.

Nelson, M. C. 1971. Classical conditioning in the blowfly (*Phormia regina*): Associative and excitatory factors. *J. Comp. Physiol. Psychol.*, **77**:353–368.

Norris, M. J. 1965. The influence of constant and changing photoperiods on imaginal diapause in the red locust (*Nomadacris septemfasciata* Serv.), *J. Insect Physiol.,* **11**:1105–1119.

Paulsen, R., and J. Schwemer. 1972. *Biochem. Biophys. Acta,* **283**:520.

Pavlov, I. P. 1927. *Conditioned Reflexes: An Investigation of the Physiological Activity of the Cerebral Cortex.* Trans. by G. V. Anrep. Dover ed., 1960, Oxford University Press, New York.

Pittendrigh, C. S., and D. H. Minis. 1971. The photoperiodic time measurement in *Pectinophora gossypiella* and its relation to the circadian systems of that species. Pages 212–250 *in* M. Menaker, ed., *Biochronometry.* National Academy of Sciences, Washington, D.C.

Prop, N. 1960. Protection against birds and parasites in some species of tenthredinid larvae. *Arch. neerl. Zool.,* **13**:380–447.

Remmert, H. 1962. Der Schlupfrhythmus der Insekten. Franz Steiner Verlag, Wiesbaden, West Germany.

Renner, M. 1957. Neue Versuche über den Zeitsinn der Honigbiene. *Z. vergl. Physiol.,* **40**:85–118.

Renner, M. 1959. Über ein weiteres Versetzungs-experiment zur Analyse des Zeitsinnes und der Sonnenorientierung der Honigbiene. *Z. vergl. Physiol.,* **42**:449–483.

Rettenmeyer, C. W. 1963. Behavioral studies of army ants. *Kansas University Science Bulletin,* **44**:281–465.

Roach, S. H., and H. R. Agee. 1972. Trap colors: Preference of alate aphids. *Environ. Entomol.,* **1**:797–798.

Roberts, S. K. de F. 1960. Circadian activity in cockroaches. I. The freerunning rhythm in steady state. *J. Cell. Comp. Physiol.,* **55**:99–110.

Roeder, K. D. 1965. Moths and ultrasound. Pages 150–158 *in* T. Eisner and E. O. Wilson, eds., *The Insects. Scientific American,* 1977. W. H. Freeman and Company, Publishers, San Francisco.

Roeder, K. D., and A. E. Treat. 1961. The detection and evasion of bats by moths. *Am. Sci.,* **49**:135–148.

Romoser, W. S. 1973. *The Science of Entomology.* Macmillan Publishing Company, Inc., New York.

Roth, L. M., and T. Eisner. 1962. Chemical defenses of arthropods. *Ann. Rev. Entomol.,* **7**:107–136.

Rothenbuler, N. 1964. Behavior genetics of nest cleaning in honey bees. 4. Responses of F, and backcross generations to disease-killed brood. *Am. Zool.,* **4**:111–123.

Ruttner, F., and H. Ruttner. 1965. Untersuchungen über die Flugaktivität und das Paarungs-verhalten der Drohnen 2. Beobachtungen an Drohnensammelplätzen. *Z. Bienenforsch.,* **8**:1–8.

Salt, G. 1970. *The Cellular Defence Reactions of Insects.* Cambridge University Press, New York.

Santschi, F. 1911. Observations et remarques critiques sur le mécanisme de l'orientation chez les fourmis. *Rev. Suisse Zool.,* **19**:303–338.

Santschi, F. 1923. L'orientation sideral de fourmis, et quelques considérations sur leurs differentes possibilités d'orientation. *Mem. Soc. Vaudoise Sci. Natur.,* **1**:137–176.

Saunders, D. S. 1976. *Insect Clocks.* Pergamon Press, Inc., Elmsford, N. Y.

Schildknecht, H., and K. Holoubek. 1961. Die Bombardierkafer und ihre Explosionschemie. V. Mitteilung über insekten Abwehrostoffe. *Angew. Chem.,* **73**(1):1–6.

Schneider, Dietrich. 1974. The sex-attractant receptor of moths. Pages 84–91 *in* T. Eisner and E. O. Wilson, eds., *The Insects. Scientific American,* 1977. W. H. Freeman and Company, Publishers, San Francisco.

Schnierla, T. C. 1953. Modifiability in insect behavior. Pages 723–747 *in* K. D. Roeder, ed., *Insect Physiology.* John Wiley & Sons, Inc., New York.

Schnierla, T. C. 1971. *Army Ants: A Study in Social Organization,* H. R. Topoff, ed. W. H. Freeman and Company, Publishers, San Francisco.

Schoonhoven, L. M. 1968. Chemosensory bases of host plant selection. *Ann. Rev. Entomol.,* **13**:115–136.

Schwartzkopff, J. 1973. Mechanoreception. Pages 273–352 *in* M. Rockstein, ed., *The Physiology of Insecta,* Vol. 2. Academic Press, Inc., New York.

Shearer, D., and R. Boch. 1965. 2-Heptanone in the mandibular gland secretion of the honey-bee. *Nature, London,* **206**:530.

Shorey, H. H. 1964. Sex pheromones of noctuid moths. II. Mating behavior of *Trichoplusia ni* (Lepidoptera:Noctuidae) with special references to the role of the sex pheromone. *Ann. Entomol. Soc. Am.,* **57**:371–377.

Shorey, H. H. 1973. Behavioral responses to insect pheromones. *Ann. Rev. Entomol.,* **18**:349–380.

Shorey, H. H., and J. J. McKelvey, Jr., eds., 1977. *Chemical Control of Insect Behavior.* Interscience, John Wiley & Sons, Inc., New York.

Slifer, E. H., J. J. Prestage, and H. W. Beams. 1959. The chemoreceptors and other sense organs on the antennal flagellum of the grasshopper (Orthoptera:Acrididae). *J. Morph.,* **105**:145–191.

Smissen, E. E., J. P. Lapidus, and S. D. Beck. 1957. Corn plant resistant factor. *J. Org. Chem.,* **22**:220.

Snodgrass, R. E. 1935. *Principles of Insect Morphology.* McGraw-Hill Book Company, New York.

Southwood, T. R. E. 1962. Migration of terrestrial arthropods in relation to habitat. *Biol. Rev.,* **37**:171–214.

Spiess, E., and B. Langer. 1964. Mating speed control by gene arrangements in *Drosophila pseudoobscura* homokaryotypes. *Proc. Nat. Acad. Sci. USA,* **51**:1015–1018.

Spieth, H. T. 1974. Courtship behavior in *Drosophila. Ann. Rev. Entomol.,* **19**:385–405.

Stephen, W. P. 1960. Artificial bee beds for the propagation of the alkali bee *Nomia melanderi. J. Econ. Entomol.,* **53**:1025–1030.

Stephen, W. P. 1962. Propagation of the leaf-cutter bee for alfalfa seed production. *Oregon State Univ. Agr. Exp. Sta. Bull.,* No. 586.

Stephen, W. P. 1973. Insects as natural resources and tools of management. Pages 31–44 *in* P. W. Geier, L. R. Clark, D. J. Anderson, and H. A. Nix, eds., *Ecol. Soc. Aust.* (*Memoirs* 1), Canberra.

Sturm, H. 1956. Die paarung beim silberfischen, *Lepisma saccharina. Z Tierpsychol.,* **13**:1–2.

Thornsteinson, A. J. 1960. Host selection in phytophagous species. *Ann. Rev. Entomol.,* **5**:193–218.

Thorpe, W. H. 1963. *Learning and Instinct in Animals,* 2nd ed. Methuen & Co. Ltd., London.

Tinbergen, N. 1951. *The Study of Instinct.* Oxford University Press, New York.

Tostowaryk, Walter, 1972. The effect of prey defense on the functional response of *Podisus modestus* (Hemiptera:Pentatomidae) to densities of the sawflies *Neodiprion swainei* and *N. pratti banksianae* (Hymenoptera:Neodiprionidae). *Can. Entomol.,* **104**:61–69.

Traynier, M. M. 1968. Sex attraction in the Mediterranean flour moth, *Anagasta kühniella:* Location of the female by the male. *Can. Entomol.,* **100**:5–10.

Ullyett, G. C. 1953. Biomathematics and insect population problems: A critical review. *Mem. Entomol. Soc. S. Afr.,* **2**:1–89.

Urquhart, F. A. 1960. *The Monarch Butterfly.* University of Toronto Press, Toronto.

VanderSar, T. J. D., and J. H. Borden, 1977. Visual orientation of *Pissodes strobi* Peck (Coleoptera:Curculionidae) in relation to host selection behavior. *Can. J. Zool.,* **55**:2042–2049.

Way, M. J. 1963. Mutualism between ants and honeydew-producing Homoptera. *Ann. Rev. Entomol.,* **8**:307–344.

Weber, N. A. 1957. Weeding as a factor in fungus culture by ants. *Anatomical Record,* **123**:638.

Wehner, R. 1976. Polarized-light navigation by insects. Pages 140–149 *in* T. Eisner and E. O. Wilson, eds., *The Insects. Scientific American,* 1977. W. H. Freeman and Co., Publishers, San Francisco.

Wehner, R., G. D. Bernard, and E. Geiger. 1975. Twisted and non-twisted rhabdoms and their significance for polarization detection in the bee. *J. Comp. Physiol.,* **104**:225–245.

Wellington, W. G. 1955. Solar heat and plane polarized light versus the light compass reaction in the orientation of insects on the ground. *Ann. Entomol. Soc. Am.,* **48**:67–76.

Wellington, W. G. 1957. Individual differences as a factor in population dynamics: The development of a problem. *Can. J. Zool.,* **35**:293–323.

Wellington, W. G. 1959. Individual differences in larvae and egg masses of the western tent caterpillar. *Can. Dep. Agr. Forest. Biol. Div. Biomed. Progr. Rep.,* **15**:3–4.

Wellington, W. G. 1960a. Qualitative changes in natural populations during changes in abundance. *Can. J. Zool.,* **38**:289–314.

Wellington, W. G. 1960b. The need for direct observations of behavior in studies of temperature effects on light reactions. *Can. Entomol.,* **92**:438–448.

Wellington, W. G. 1964. Qualitative changes in populations in unstable environments. *Can. Entomol.,* **96**:436–451.

Wellington, W. G. 1965a. Some maternal influences on progeny quality in the western tent caterpillar *Malacosoma pluviale* (Dyar). *Can. Entomol.,* **97**:1–14.

Wellington, W. G. 1965b. The use of cloud patterns to outline areas with different climates during population studies. *Can. Entomol.,* **97**:617–631.

Wellington, W. G. 1974a. A special light to steer by. *Natur. Hist.,* **4**:47–52.

Wellington, W. G. 1974b. Bumblebee ocelli and navigation at dusk. *Science*, **183**:550–551.

Wellington, W. G. 1976. Applying behavioral studies in entomological problems. Pages 87–97 *in* Anderson and Kaya, eds., *Perspectives in Forest Entomology*. Academic Press, Inc., New York.

Wellington, W. G. 1977. Returning the insect to insect ecology. Some consequences for pest management. *Environ. Entomol.*, **6**:1–8.

Wellington, W. G., C. R. Sullivan, and G. W. Green. 1951. Polarized light and body temperature level as orientation factors in the light reactions of some hymenopterous and lepidopterous larvae. *Can. J. Zool.*, **29**:339–351.

Wenner, A. M., and D. L. Johnson. 1967. [Reply to K. von Frisch. *Science*, **158**:1072–1076.] *Science*, **158**:1076–1077.

Wenner, A. M., P. H. Wells, and D. L. Johnson. 1969. Honeybee recruitment to food sources: Olfaction or language? *Science*, **164**:84–86.

Wheeler, W. M. 1910. *Ants: Their Structure Development and Behavior*. Columbia University Press, New York.

Whiting, P. W. 1932. Reproductive reactions of sex mosaics of a parasitic wasp. *J. Comp. Psychol.*, **14**:345–363.

Wickler, W. 1968. *Mimicry*. Trans. by R. D. Martin. World University Library. McGraw-Hill Book Company, New York.

Wigglesworth, V. B. 1956. *The Principles of Insect Physiology*. Methuen & Co. Ltd., London.

Wigglesworth, V. B. 1964. *The Life of Insects*. The New American Library, Inc., New York.

Wigglesworth, V. B. 1965. *The Principles of Insect Physiology*, 6th ed. Methuen & Co., Ltd., London.

Williams, C. B. 1970. The migrations of the painted lady butterfly *Vanessa cardui* (Nymphalidae), with special reference to North America. *J. Lep. Soc.*, **24**:157–175.

Wilson, E. O. 1962. Chemical communication among workers of the fire ant *Solenopsis saevissima* (Fr. Smith). 1. The organization of mass foraging 2. An information analysis of the odour trail. 3. The experimental introduction of social responses. *Anim. Behav.*, **10**:134–164.

Wilson, E. O. 1963. Pheromones. Pages 92–101 *in* T. Eisner and E. O. Wilson, eds., *The Insects. Scientific American*, 1977. W. H. Freeman and Company, Publishers, San Francisco.

Wilson, E. O. 1971. *The Insect Societies*. The Belknap Press of Harvard, University Press, Cambridge, Mass.

Wilson, E. O. 1975. Slavery in ants. Pages 257–262 *in* T. Eisner and E. O. Wilson, eds., *The Insects. Scientific American*, 1977. W. H. Freeman and Company, Publishers, San Francisco.

Woods, E. F. 1959. Electronic prediction of swarming in bees. *Nature*, **184**:842–844.

Wratten, S. D. 1976. Searching by *Adalia bipunctata* (L.). (Coleoptera:Coccinellidae) and escape behaviour of its aphid and cicadellid prey on lime (*Tilia vulgaris* Hayne). *Ecol. Entomol.*, **1**:139–142.

Yokohari, F. 1978. Hydroreceptor mechanism in the antenna of the cockroach *Periplaneta. J. Comp. Physiol.*, **124**:53–60.

Index

Boldface numerals indicate definition or major treatment; *italics* indicate illustration; "t" indicates table.